A Blessing
of Toads

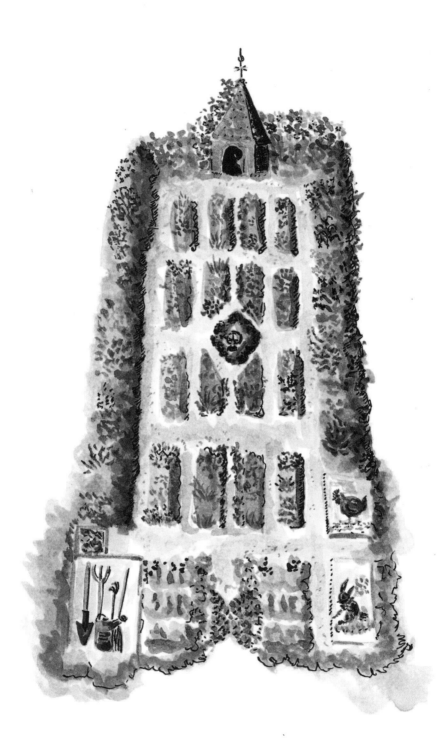

A Blessing of Toads

A Guide to
Living with Nature

Written & Illustrated by
SHARON LOVEJOY

Down East Books

Camden, Maine

This book is dedicated to two generations of loved ones:
Kate Pemberton Stearns
Who shared friendship, laughter, wisdom
and the joy of full moons shining across John's Bay

And to my beautiful granddaughter Sara May Arnold
who I hope will grow to love all the simplest pleasures of nature

欣
兒

Published by Down East Books
A wholly owned subsidiary of The Rowman & Littlefield Publishing Group, Inc.
4501 Forbes Boulevard, Suite 200, Lanham, Maryland 20706
www.rowman.com

Unit A, Whitacre Mews, 26-34 Stannery Street, London SE11 4AB

Distributed by NATIONAL BOOK NETWORK

Copyright © 2015 by Sharon Lovejoy
First paperback edition 2015

Cloth edition originally published in 2004 by Hearst Books, a division of Sterling
Publishing Co., Inc.

All essays in this collection were previously published in *Country Living Gardener*.

Book design by Celia Fuller.

British Library Cataloguing in Publication Information Available

Library of Congress Cataloging-in-Publication Data

The hardback edition of this book was previously cataloged by the Library of
Congress as follows:

Lovejoy, Sharon, 1945–
   Country living gardener : a blessing of toads : a gardener's guide to living with
   nature / Sharon Lovejoy
      p. cm.
      1. Gardening 2. Gardens 3. Garden animals 4. Lovejoy, Sharon, 1945– I. Title.
   SB455.L655 2004
   591.75'5–dc22

                                                                          2003021000

ISBN: 978-1-58816-379-2 (cloth : alk. paper)
ISBN: 978-1-60893-359-4 (pbk. : alk. paper)
ISBN: 978-1-60893-360-0 (electronic)

♾™ The paper used in this publication meets the minimum requirements of
American National Standard for Information Sciences—Permanence of
Paper for Printed Library Materials, ANSI/NISO Z39.48-1992.

Printed in the United States of America

CONTENTS

FOREWORD

Too often, the voice of experience comes to us in the form of a haughty scold or a boring dissertation: Not so in the case of author Sharon Lovejoy. Playful, modest, and full of fun, Sharon's lighthearted essays and whimsical watercolor illustrations combine scientific fact with scholarly observation in a manner that captures our hearts and makes us wiser as if by magic.

Rooted in memory and nourished by season after season of hands-on horticultural experimentation, Sharon's common-sense tales reflect an artist's eye and a poet's sensibility. Here is a world in which every walk in the woods, summer by the sea, or hour in the garden promises us an opportunity to understand something new about nature—and about ourselves. And here is a gardener who knows flowers by their Latin names and what thrushes prefer for breakfast. She is a planter of trees, a lover of ponds, and a defender of wasps, bats, and toads—and her honest enthusiasm for many of the garden's most maligned creatures effectively convinces us to rally to their defense, too. Interpreted by Sharon's imagination, moles, squirrels, honey-bees, skunks, beetles, snakes, snails, and plants ranging from the smallest weed to the tallest tree have more per-sonality—and more staying power—

than the characters you'll find chronicled in any of this week's celebrity tabloids or gossip columns.

For more than ten years, the readers of *Country Living Gardener* have come to Sharon's "Heart's Ease" column to be entertained. And they have left feeling educated and befriended. For those devoted readers—and for gardeners, backyard biologists, and kindred spirits everywhere—we offer this collection of essays, illustrations, and weekend projects. Let them help you to rediscover the poetry, beauty, charm—and fun—of the world around you. The characters you are about to meet will stay with you your whole life.

MARJORIE GAGE, EDITOR
Country Living Gardener

ACKNOWLEDGMENTS

I am grateful to a handful of exceptional people who helped me through my years of writing. My thanks go to Niña Williams, who encouraged me from our first meeting. Special thanks to Diana Gold Murphy for believing in my work and introducing it to Hearst Books. A bouquet of *mercis* to the adorable Alexa Barre of *Country Living Gardener*, who keeps me in line, to Ruth Rogers Clausen, the best horticultural editor anyone could have, to the creative talents of Celia Fuller, Design Director, and to my sensitive and encouraging editor, Lesley Bruynesteyn, and the rest of the staff at Hearst Books. And, always to my son Noah Arnold for his loyalty, and to my husband, Jeff Prostovich, who helps me every step along the garden pathway.

INTRODUCTION

The memories are distant and nebulous. They flash in and out of my consciousness like the glimmer of fireflies captured in a Mason jar. Sometimes just the scent of old-fashioned pinks or the sight of an iridescent dragonfly will conjure those memories and the images and enchantment of my grandmother's garden where I spent the first seven years of my life. Grandmother Lovejoy's riotous garden wasn't perfectly groomed, nor did it have a planned palette of matching colors or an architecturally designed hardscape. What it did have was LIFE, a constant humming, busyness, and rhythm that made me want to leap out of bed each morning and head outdoors for an adventure, a feeling I have to this day.

Many years ago, I realized that I could make a career of sharing my passion for gardening and nature. In 1982 I founded Heart's Ease Herb Shop & Gardens where I taught classes, led natural history walks, and held festivals. In 1989 *Country Living* magazine sent editor-at-large Niña Williams to do a feature article about my business, gardens, and California cottage "Seekhaven." As Niña and I became better acquainted, she told me that she had a dream of starting a gardening magazine for *Country Living* and that she wanted me to write for her. I couldn't believe my good fortune. A few years later, Niña called to tell me that *Country Living Gardener* magazine would be a reality, and true to her word, she invited me to write a garden and nature column named after my business. In March

1993 I wrote my first "Heart's Ease" column, which wasn't about how to plant a garden but how to appreciate it and its myriad inhabitants. To this day, that same theme is the thread of continuity in every issue, whether I write from my California gardens or my seaside cottage in Maine.

My columns don't come easily. Sometimes I spend years observing my subject before I set my pen to hand for an illustration or a word, practicing what the poet Wordsworth called "the harvest of a quiet eye." To observe something in its own environment, to learn how it affects the health and productivity of plants and the surrounding landscape, and to discover the little mysteries of each garden critter or plant are what really intrigue me and what I want to share. Yes, I love flowers, trees, shrubs, grasses, and herbs, but what I love most is the life that they engender and that, in turn, engenders them.

One of the rewards of writing is the response from readers who look at their garden and its dwellers in a new and more gentle way. Nothing makes me feel better than to learn that a gardener is no longer trapping moles or that someone is now unafraid of bumblebees, snakes, or other beneficial critters found in their yard. Now my readers have rewarded me in another way. Because of their requests, the editors at Hearst Books have combined my previous columns into this book.

I hope A Blessing of Toads, which is my heartfelt offering to you, will encourage you to leap out of your bed each morning to enjoy your garden. Perhaps it will help you to view the mysteries of the earth as you did in childhood, when everything was fresh and exciting. This isn't a journey backward, but a journey forward, closing the circle in a completeness found only in the simplest of joys and pleasures.

A century ago, the naturalist John Burroughs wrote, "I still find each day to be too short for all the thoughts I want to think, all the walks I want to take, all the books I want to read, and all the friends I want to see." I agree with Burroughs, but I need to add another sentence to his ruminations on time. I still find each day to be too short for all the gardens I want to plant, all the critters I want to know, all the mysteries I want to explore, and all the friends with whom I wish to share them.

Carpe terra!

SHARON LOVEJOY

A Feast for the Eyes

Common names for Viola tricolor are love-in-idleness, Johnny-jump-up, and heartsease; by any name, this enchanting flower symbolizes the cheer a garden can give.

When I was a child of twelve, I often sat with pen and paper in the boughs of a favorite giant sycamore, quietly looking down as my neighbors harvested fruits and vegetables from their garden. As I mused, I wrote over and over, "Seekhaven," the name of my dream home, a home that would provide heartsease and shelter to all who entered. Seekhaven's garden, my own garden, would have more than plants; birds, butterflies, and animals would be welcome, too. I dreamed of gathering harvests as rich and bountiful as the one my neighbors were gathering below me.

My friends thought I was crazy. They were busy reading mystery stories and teen magazines, and I was reading books by Gene Stratton Porter, memorizing Wordsworth's poetry, devouring Mother's home magazines, and dreaming of my own first crop of apples.

Today, I am home. Tucked away under ancient oaks and pines and a tall, shaggy-barked cedar tree is an old barn-red cottage trimmed in green. A carved and painted sign swings under an arbor thick with morning glories and jasmine. The sign reads simply "SEEKHAVEN."

This morning I awake with a smile on my face, and the cheerful faces of heartsease overflowing from my window box smile back at me. The mingled patter of gentle rain and a covey of quail skitter across the leaf-covered rooftop.

Despite the rain, my morning ritual of garden patrol must go on. As a gardening friend, Ruth Pearson, says, "You must walk between the raindrops." This is the time of day when my flannel nightgown drags across the moist, rich soil as I weed, dream, plan, and dig tentatively with an old silver spoon. The quail have moved from their rooftop and are now clucking and making a sound peculiarly like Grandmother Nonie's old agateware coffeepot just as it would start to perk. They are letting me know the weeds can wait, but they must be fed.

I scoop out mounds of black sunflower seeds for the jays and chickadees sitting on peach branches strung with glittering raindrops. Cups of cracked corn and seed are flung out to the quail as the tame squirrel tries to open the top of the storage container.

With the needs of my "children" met, I continue a quiet garden survey. Veils of gentle rain have intensified the myriad odors of the herb garden. Spears of emerging daffodils punctuate the edges of all the trails, and a haze of fresh green leaves laces the gray arms of the monarch butterflies' willow roost.

Just as I caress the fragrant, papery leaves of lemon verbena, a Lesser goldfinch bursts through the branches with an alarmed chitter. Cradled in the twigs of the verbena is one of nature's mysteries—a small, cup-shaped nest of fine grasses and some white hair from my patient shepherd dog, Una. A brilliant sapphire-blue feather from a Steller's jay is jauntily angled into the weaving as if it were the crowning touch of a skilled milliner. Inside the nest, three sky-blue eggs sit shouldered together on a patch of green moss and lichen. It's no wonder the goldfinch greeted me less than graciously.

Where has the morning gone? As I work my way back toward the house, I snip at herbs and early butter lettuce for a fresh salad. Along the stone pathway, murmuring quail graze amongst colorful patches of heartsease—the birds are harvesting their own salad! I am smiling again.

A garden can give many harvests, but perhaps the most important is the one that awakens our spirits every single day. Wordsworth described it as the "harvest of a quiet eye." Billowing bouquets of apple blossoms, the lilting flight of a cobalt dragonfly, giant baskets of golden pumpkins, and the tiniest hummingbird bathing in a dew-filled leaf are just part of the feast. Harvest all manner of a garden's offerings, and you will have learned the heartsease your own garden can provide.

The Seediest Garden in the Neighborhood

And proud of it!

My day began with noble intentions. I planned to clip and clean my winter gardens until they were pristine enough to receive a visit from Rosemary Verey, the celebrated British gardener. Instead, I sat motionless on the ground and watched as a green-backed Lesser goldfinch swooped down and perched on the dried cosmos I had mounded in my harvest basket. For five minutes, the small bird picked through my pile of spent flowers like a bargain hunter at a flea market and then flew to the top of the only cosmos still standing in the garden. For the next few minutes, the goldfinch held on to the swaying plant and teased out seeds. The thought flashed through my mind that perhaps I was being too finicky about neatness. Maybe Rosemary wouldn't notice my bedraggled plants.

The line of sunflowers around the playhouse looked miserable and needed immediate grooming. Their seed-heavy heads, once golden and humming with bees, now drooped like forlorn children. A chestnut-backed chickadee, dwarfed by the tall flowers, hung upside down and tugged out plump, dark seeds while a Steller's jay complained loudly at the chickadee's pilfering. The tidying of the tattered sunflowers would have to wait until the birds finished their arguments.

I picked up my basket and moved into the children's garden, where I intended to clean out a giant terra-cotta pot of broom corn (Sorghum spp.) that had seen better days. I stopped under the arbor and re-evaluated my plans. The pot was occupied by white-crowned sparrows who were so busy eating that they didn't even notice me as I walked past them toward the butterfly garden.

The goldfinch was there ahead of me. He ignored the sign that stated how much the butterflies enjoy sipping nectar from the mini-meadow of annual and perennial flowers. His attention was focused on the seeds hidden in the showy, cone-like heads of purple cornflowers and the remnants of a patch of coreopsis and rudbeckia long gone by. His patient examination and probing of each haggard flowerhead yielded enough tidbits to keep him busy for nearly half an hour. While he was busy, I was not.

It occurred to me that the gardening plans I formulated the night before were just not going to happen. My cleanup day turned into a practical lesson on sharing a garden with the birds. I learned in a few hours of quiet observation how much they depend on a cargo of summer seeds to carry them through the insect-poor months of winter.

During my lunch break, I pulled some reference books off the shelf and made a list of cultivated plants favored by birds. Topping the chart was corn, which feeds over eighty-one species of birds as well as numerous animals, including beaver, fox, muskrat, raccoon, skunk, squirrel, opossum, and mice. I've had firsthand experience with corn's amazing popularity—my "Three Sisters of Life" garden was nibbled by a parade of critters who seemed to enjoy it in every stage of growth, from its first needlelike sprout to its eventual resting place in a compost heap, where it was picked clean by a family of crows.

Just a few points below corn on the popularity chart is wheat, which appeals to seventy-four species of birds and twenty animals. I was trying to figure out exactly which birds eat the most wheat and noted that 53 percent of the diet of the snow bunting is composed of wheat, that blue grosbeaks favor it heavily, and that the three-foot-tall sandhill crane may devote 50 percent of his intake to wheat. A 1926 publication by ornithologist Arthur Cleveland Bent stated that although the crane is opportunistic and eats snakes, frogs, lizards, and insects, during fall migration it subsists on wheat, which it gleans from the stubble fields along its route.

My friend John Arnold of Albuquerque is a gardener extraordinaire and a crane lover as well. Every fall, he calls to tell me of the flocks of trumpeting sandhill cranes flying over his gardens on their way to their wintering grounds in Bosque del Apache, New Mexico. I'm going to drop John a note and give him a gardening hint that may seduce those cranes down from the sky and into his yard. Plant wheat.

My lunch hour turned into an afternoon as I piled up more books and added plants to my list. Millet is the choice of about sixty species of birds; sunflowers satisfy fifty; barley appeals to about forty-two species; and sorghum trails with thiry-two.

I read through an old Seeds Blum catalog which offered broom corn as well as Mennonite, White African, and Brandese sorghum, and claimed, "No bird seed mix worth its suet can afford to be without sorghum seeds." I remembered the white-crowned sparrows stuffing themselves on my pot of sorghum and had to agree.

Shepherd's Garden Seeds' newest catalog has a "Songbird Collection," which features painted calliopsis (Coreopsis tinctoria), birdspray millet (Pannicum spp.), golden 'Peredovik' sunflowers (Helianthus annuus), and golden orange safflower (Carthamus tinctorius). The staff at Shepherd's did their homework—they point out that birds also like to dine on the seeds of black-eyed Susan, purple coneflowers, tithonia, and love-lies-bleeding. They end their garden design hints with these words of wisdom: "When you view your yard as an ecosystem, magic happens and wildlife moves in."

John and Ruth Lopez of Point Reyes, California, have a family business that promotes the garden as a unique learning environment. Their catalog, Gardens for Growing People, offers chemical-free "Living History Seed Collections" created especially for them by Kids in Bloom of Zionsville, Indiana. The Lopez's catalog highlights a "garden for the birds," which includes giant sunflowers, coreopsis, sweet rocket, towering red Uhuru amaranth, and a

• For the Birds •

The birds need your help during the cold winter months. Always provide them with a supply of fresh, clean water and...

- Save cuttings of wheat, corn, sorghum, millet, and sunflower heads to fashion into wreaths, swags, and sheaves that perform double duty as outdoor decorations and bird fodder.
- Decorate a tree with strings of popcorn, peanuts, and cranberries. Dangle cookies and cake donuts from the branches. Cut oranges and apples in half and fasten them onto limbs.
- Transform an ordinary snowman into a birdman by costuming him in colorful fruits, vegetables, nuts, and popcorn.
- Stuff pinecones with a mixture of suet, raisins, sunlower seeds, peanut butter, flour, and cornmeal. Tie with string or wire and suspend from a beam or railing out of reach of cats.

packet of gourd seeds, so kids can grow their own gourd birdhouses. Their cultivation requirements are written in child-friendly terminology, and there is a wealth of cultural and historical information included in each package.

My catalogs from Burpee, The Cook's Garden, and Seeds of Change all offer an incredible array of sunflowers sure to please even the pickiest of birds. One of the newest varieties, 'Paul Bunyan', was developed exclusively for Burpee by a researcher who has been breeding sunflowers for twenty years. The breeder's new offspring grows to a whopping fifteen feet. Just one of those giants would feed an entire family of hungry house finches.

As I finished my sandwich and tea and stacked up catalogs and reference books, I ran across my well-worn copy of Canon Ellacombe's nineteenth-century classic, In a Gloucestershire Garden. Ellacombe wrote, "A garden without birds is like a garden without flowers." I realize that for me the pleasures of a garden are inextricably bound to my interactions with birds. I can't imagine a garden bereft of birds or flowers. One without the other would be only half an experience, as incomplete as a silent concert hall or a childless playground.

Although it is winter, I am going to use all of the information I gleaned today to plan for a new garden rich with seed-bearing plants. Instead of just a pot of sorghum, I'll plant a broad swath and include a mixture of wheat, millet, and corn. I'll cultivate a brilliant labyrinth of sunflowers and an arbor of pendulous birdhouse gourds, and sow the minimeadow with a dazzling display of flowers, for the butterflies and the birds.

If I feel I must do some cleanup, I'll adopt the Norwegian tradition of binding my harvest into sheaves, and hang them on my porch as a thank-you to the birds. But, from now on, I'll leave the unruly cosmos standing, in honor of the goldfinch who changed my plans when he flew in for a morning visit and a snack.

Pond Parenthood

Discover the pitfalls, pleasures,
and possibilities of a water garden.

I have always dreamed of having a pond in my backyard. I worked toward my dream by water gardening in a homely, shallow cement pool, sunken galvanized buckets, antique crocks, giant half barrels, and, finally, a real pond I dug myself. Though I love my pond as a parent unconditionally loves a child, I realize that it is perhaps the ugliest one I have ever seen.

My earliest water-garden experience was on the banks of the lily pond at the Huntington Library in San Marino, California, where I wandered and played for hours. That serene, lily-starred pool, with its chorus of frogs and birds and squadrons of patrolling jewel-colored dragonflies, inspired my first attempt at creating a pond.

During the summer of my ninth year, I devoted an entire weekend to digging a four- by three-foot kidney-shaped pool at the back of my parents' suburban yard. Soil as hard as cement made the shoveling torturous. By the time I had dug as deeply as any nine-year-old could be expected to, I had excavated just a few inches, which I assured myself was sufficient.

I dragged our long hose to the edge of my creation and slowly began adding water. I was excited and full of dreams of frogs and toads and my own company of visiting dragonflies. After an hour of running a slow, steady stream of water, my pond was full, but closely resembled a large helping of chocolate pudding. I remember thinking that it would be clear and sparkling by the morning. The next day when I awoke and ran outside, I faced a soggy, empty hole that was only suitable for supplying mud to a troop of barn swallows busily patching their sagging nests.

My teenage years arrived and water gardening was the furthest thing from my mind. I didn't return to that passionate interest until I inherited a shallow water-filled cement pool at the edge of Tecolote Canyon in San Diego. That pond, so devoid of personality and style, acted like a wildlife magnet. Dozens of species of birds swooped in for drinks and a bath. Toads and frogs laced the surface of the water with long, clear strands and masses of jelly, peppered with eggs. The shy fox who haunted the edges of the canyon crept into the garden every morning and warily drank his fill.

Leaving a garden is always sad, but I was doubly saddened when I walked away from my small wildlife habitat and moved into an apartment. Forced to limit my garden to a crowded, walled patio,

I transferred my passion for water gardens to a succession of containers that turned into a parade of stinking failures clogged with algae and dying fish and plants.

One afternoon, I curled up with a copy of King Solomon's Ring by noted naturalist Konrad Lorenz. Lorenz's writings stress animals and plants living together in biological equilibrium. By providing at least 60 percent coverage of the pond surface with aquatic plants, a few small fish for mosquito control, and snails for cleanup, I would be able to achieve a balanced ecosystem. I successfully applied his theory to water gardens—first in a galvanized bucket, then in a twenty-gallon antique crock, and finally in a giant half wine barrel, which is still thriving and healthy.

I have visited dozens of exciting water gardens in the past few years, and the diverse ponds reawakened my only slightly submerged desire for one. In Lockport, New York, Bud and Rita Daubins' pond charmed me for hours with its burgeoning population of bluets (damselflies) that laced the reflection of the sky to the glistening water. In the parched landscape of Albuquerque, New Mexico, John Arnold's small pond bustled with the zig-zagging antics of black-and-white-striped dragonflies darting like frantic referees at a basketball game. From the sidelines, under a terracotta pot, a secretive Woodhouse's toad looked on. In the walled "Family" area of the busy New York Botanical Garden, Catherine Eberbach has created a small pond habitat that entices and captivates not only wildlife, but also the nature-starved children of the inner city.

I returned home from my travels inspired and ready to begin planning for my longed-for pond. Early one morning, I abandoned all rational thought, uncoiled an old hose to outline my proposed pool, and started digging. After a few hours of work, I recollected a page of cautionary notes in the Lilypons Water Gardens catalog. Instead of stopping and reviewing the information, I kept working until I couldn't lift another shovelful of soil.

The next morning I hobbled into the garden and assessed the small hole I had dug. I realized that this pond-digging business was going to take a lot more work and time than I had thought. I made a solemn pledge to dig for thirty minutes every morning until the pond was the proper depth and shape. After a long, hard month, I walked outside to my eight- by twelve-foot hole in the ground and decided I was finally finished with digging.

My friends Barbara and Bob Brookins came to visit and saw my "finished" excavation. Shaking their heads in dismay at the mess I had created, they jumped in and started refining, leveling and smoothing the walls of the hole until it was ready for the next step.

The following week my living room was carpeted with catalogs as I tried to choose a suitable liner. Should it be EPDM fish-grade rubber, PVC liner, or the lighter, more flexible, and expensive Butyl? Should I cement it or purchase a preformed fiberglass shape? Did I need an underlayment or should I opt for a bonded geo-textile material? When I began, I thought that digging my dream pond would be the hardest chore, but there didn't seem to be *any* step that was simple. Finally I chose a fourteen- by twelve-foot EPDM 45 mil liner with an underlayment of a long-lasting synthetic material.

My husband Jeff and I spent a morning unpacking and unkinking the heavy liner and underlayment. We worked patiently until the pond was draped and fitted with the unmanageable black material. I groaned when we finished: The pond looked smaller and closely resembled the homely kidney-shaped pool of my childhood.

• How to Build a Pond •

Follow These Simple Guidelines

- Select a flat site away from overhanging trees and shrubs.
- Outline your proposed pond with string or a garden hose.
- Dig pond to a depth of 3 to 4 feet, and slope your walls slightly inward at a 70-degree angle (this keeps raccoons from invading).
- Purchase a 45 mil EPDM pond liner, which is guaranteed to last twenty years.
- Spread liner evenly in pool, tucking and folding along edge as necessary. Cover edges with boulders or flat rocks.
- Create a cavern area in the center of the pond with rocks or bricks (a hiding place for pond dwellers).
- Fill with water and add a de-chlorinator.
- Let pond sit for two weeks before adding fish.

It began to rain and the runoff from the slope behind us turned into small brown rivulets filling the liner and trickling underneath it. Jeff grabbed a shovel and dug a trench to redirect the flow. I ran to the lower terrace and hoisted buckets of loose soil to refill the eroding edges. As the pounding rain and wind began to lift the sides of the liner, Jeff pried up rock stepping stones to edge the pond and keep the liner in place. After a full day of work, we had disassembled our walkway to create an edging for something that looked like an illustration of how NOT to make a pond.

"Hope springs eternal," wrote Alexander Pope, and I must admit

- Visit a nursery or peruse water garden mail order catalogs for plants. Select from an array including water lilies, but be sure to add some floaters, such as floating four leaf clovers (*Marsilea quadrifolia*), and submerged plants, which are the water purifiers.
- Purchase some water snails, the vacuums of the pond.
- Keep 75 percent of the water surface covered with plants.
- If you want to encourage frogs and toads, DON'T fill your pond with large Koi or goldfish, which will eat tadpoles. Try the tiny, native *Gambusia affinis* that are known as "topsiders." They'll scour the upper reaches of your pond for mosquito larvae, and they will never need supplemental feeding.
- To keep your pond healthy, once a month you may want to add a microbe-lifter, available at your local nursery or catalog supplier.

that when I filled the pool with water, my hopes soared. It almost looked good. I knew that once it was stocked with plants and fish it would be perfect. I drove to Sheltered Acre Nursery in Los Osos, California, to confer with Lowell, the knowledgeable proprietor. He walked patiently from tank to pool, lifting dripping plants for my approval. I chose floating four-leaf clovers (*Marsilea quadrifolia*), dwarf papyrus (*Cyperus* spp.), a regal pink water lily (*Nymphaea* 'Marliac Carnea'), floating water lettuce (*Pistia stratoites*), water primrose (*Ludwigia arcuata*), and an array of smaller plants that would perform well despite my shady conditions.

Lowell wrapped the plants in plastic bags, then added water to another bag and deftly netted two dozen gambusias (*Gambusia affinis*), small North American top-feeding fish nicknamed "mosquito fish" for their ability to keep ponds free of pesky mosquito larvae. Next, he rolled up his sleeves and plunged his hands into a tank thick with a tangle of submerged greenery. The plants he uprooted— eel grass (*Vallisneria* spp.), milfoil (*Myriophyllum* spp.), waterweed (*Elodea* spp.), and cabomba (*Cabomba* spp.)—are not known for their showiness but for their ability to function as hardworking pond purifiers. Finally, he plucked eighteen water snails, "pond vacuums," from the sides of a tank and dropped them into one of my bags.

I drove home and gently carried my treasures to the pond. I floated the fish bag, unwrapped the plants, slipped out of my blue jeans, and waded up to my thighs in cold, clear water. By placing blocks and bricks in the bottom of the pool, I created varying depths to suit the needs of all of the plants. Next time, I thought, shivering with cold, I'll build a pond with ledges and shelves instead of hauling around backbreaking blocks. I climbed out, pulled on my jeans, and smiled with satisfaction.

The next morning, I walked out with coffee cup in hand to survey my new creation. The once-clear pool was darker than my drink, muddy and strewn with long strands of shredded water lily. Empty plastic pots floated upside down, one bent rush pierced the surface of the water, and most of the snail shells were sucked clean. My wildlife magnet had attracted its first visitors, hungry raccoons.

I made a frantic call to Lowell and asked for advice. His first helpful hint was that I should have dug my pond with walls sloping slightly inward at about a 70- to 75-degree angle. He mentioned that the water should be at least two-and-a-half to four-feet deep to discourage raccoons, who evidently don't like swimming for their supper. He suggested building caverns of rocks where fish and frogs can hide, and he advised me to move my largest, most expensive plants into the deepest water, far from the eager grasp of the raccoons. Lowell said that some people have successfully deterred raccoons by lacing thin nylon thread across and alongside their ponds. I opted to wade in again, rearrange the blocks and bricks in the deepest area, create a cavern, and tuck the tattered survivors back into their pots and set them in their new places.

The raccoons have given up their midnight raids, and living with a pond has settled into an easy rhythm. Every day brings new discoveries. Water-strider spiders skim the surface on dimpled footprints while underwater dragonfly naiads and backswimmer bugs vie for prey. Needle-like sapphire damselflies curl their abdomens into question marks as they deposit eggs in strings of algae. A procession of birds and butterflies sunbathes on the flat rocks and drinks at the pond's edge. The gosling-like peeping of a hidden Western toad accompanies a fervid opera of the amorous frogs that serenade me through the warm summer nights.

I put the pond in the wrong place. I didn't dig to the proper depth. The shape is boring and the stone edging looked better as a pathway. I have learned from my mistakes and will do things differently next time, but my unconditional love has paid off. Despite its ugliness and problems my pond is healthy, thriving, and a constant source of joy and wonder. No parent could ask for more.

Windowsill
Reveries

*Pass-throughs to nature nurture many
personal plant collections.*

As the days shorten and the sun rides lower in the sky, the windows of my old California cottage are flooded with rich, golden light. My workroom, a bright glassed-in porch overlooking gardens and the distant Santa Lucia Mountains, is bordered by broad wooden sills filled with a collection of plants.

My love for windowsill gardens is firmly rooted in memories of childhood. The wide, sunny sills in my grandmother's kitchen and living room were always crowded with plants in various stages of exuberant growth. No horticultural connoisseurship was being flaunted, just a perennially green thumb that wanted to give life to every root and shoot passing over the threshold.

As a child I watched as single, velvety African violet leaves in platoons of tiny medicine vials put out thready roots and were then graduated into cracked teacups filled with soil. Carrot-top forests flourished in blue willow saucers and lunchbox leftovers produced a mini orchard of apple, orange, avocado, and lemon trees. Yams, so despised on my dinner plate, were poked with a necklace of tooth-picks, suspended over Mason jars full of water, and moved into positions of honor on the windowsills flanking the fireplace. The homely yams sprouted, sent out a few timid shoots, and then, almost overnight, galloped across the mantel, and twined and looped over the curtain rods, across the china cabinet, piano, and bookcases, and back into the kitchen where their journey began.

Lily of the valley pips (roots) were nestled into moist sphagnum moss and tucked into a celadon bowl. Hyacinth bulbs moved from the refrigerator to a medley of narrow-mouthed vases occupying one whole shelf of the dark, cool linen closet. When their rubbery white roots filled the vases and their pale green leaves had grown a sturdy six inches, they were moved onto a bright windowsill grudg-ingly shared by Caesar, Grandmother's ill-tempered marmalade cat. Within weeks Caesar would be surrounded by a bevy of purple, blue, pink, and white hyacinths cloaked in fragrance.

I remember skating along the sidewalks in my neighborhood and noticing what other people were nurturing on their win-dowsills. Mrs. Braden had a rainbow of begonias peering through her glass and Mr. Ross paraded legions of sprouting vegetables on all of his south-facing sills. I was hooked...nobody's windowsill would go unnoticed.

Decades have passed, and I am still a shameless gawker and windowsill snoop. Old sketchbooks overflow with notes and drawings of my favorite windowsill gardens and with ideas for my own indoor gardening experiments. In Devon, England, I stopped to draw a sunny bay window and discovered the unexpected sight of 'Heavenly Blue' morning glories growing in a terra-cotta pot indoors. The vines climbed a twenty-four-inch tall willow tepee, and their cerulean blooms glimmered through the wavy panes. In Sundborn, Sweden, I sketched a wide marble-topped windowsill that harbored legions of brilliant geraniums, a ninety-year-old Christmas cactus, and a clay pot crowned by a ring of moss and covered with strawberries that dangled and glistened like ruby earrings.

Through the years I have mothered a motley crew of notable plants. Scented pelargoniums, one of the most popular houseplants in

Victorian times, remain my personal favorites. I use their aromatic leaves in potpourri and tuck them inside letters, cook with their flowers, and simply enjoy brushing past them to release their pungent scents. In autumn I always have a refrigerator full of hyacinths, and a copper tray filled with gravel and the promise of paper-white narcissus. This past year found me yearning to go farther afield ... to experiment with and experience a wider variety of indoor plants.

For guidance, I turned to Sue and Joseph Brungs of The Old Greenhouse in Cincinnati, Ohio, and the sprightly Tovah Martin, horticulturist at Logee's Greenhouses in Danielson, Connecticut, and author of the informative and invaluable book *Well-Clad Windowsills*.

Sue and Joseph's kitchen has lots of light and enough humidity to successfully grow standards of rosemary and myrtle, a Monterey cypress clipped like a Christmas tree, a narrow-leafed Ficus, a good selection of herbs, and my favorites—an array of mini, animal-shaped Serissa topiaries clipped and groomed to perfection.

Tovah Martin's windowsills and the pages of her book are both overflowing with myriad plants—everything from abutilon to zygocactus. Since Tovah can't answer all of my gardening questions personally, I simply pick up my tattered, water-stained copy of her book and quickly find an answer to any problem that arises. From this volume, I also have gleaned a bushel-basketful of historical information: "Herbs were the first houseplants, welcomed indoors because they had a service to perform.... English lavender was once placed in the windows of wayside taverns to signify that the linens were clean.... NASA suggests growing houseplants to purify the atmosphere indoors." Tovah's lively and engaging words keep me hopping through the pages and making lists of the plants I am going to try to grow this winter.

My mailbox is brimming with seed catalogs. The refrigerator is crammed to overflowing with bulbs. A circlet of 'Pink Panda' strawberries jostles for window space with a small tepee of 'Heavenly Blue' morning glory seedlings and a copper tray of gangly narcissus. Like millions of gardeners everywhere I am an optimist...looking toward spring, galloping toward the light like my grandmother's rambunctious, homely windowsill yams.

A Member of the Passalong Club

"People don't own the wonders of nature, they just take care of them for a time. What brings joy to one should bring joy to all."
—STEVE BENDER & FELDER RUSHING,
Passalong Plants, 1993

Gardening friends taught me the meaning of the old phrase "generous to a fault." My garden is mute testimony to that generosity. Though I've tried during the past fourteen years to introduce only natives to my oak-shaded California garden, it just hasn't worked out the way I so carefully planned, thanks to the Passalong Club.

How could I refuse the moonwort, or silver dollar plant (*Lunaria annua*), that my neighbor Louise Squibb assured me would be "right at home in the garden"? It is not only right at home, it has claimed dominion over shady nooks and sunny spots and cracks in our old rock and concrete pathways. I've never met a plant with such a will to live.

Moonwort is one of the classics that authors Felder Rushing and Steve Bender call a "passalong plant." They explain: "Old-fashioned plants and passalongs are not necessarily the same. For example, a plant may have been around for ages and evoke fond memories, but if it's difficult to propagate, it's unlikely to be handed from neighbor to neighbor." I can personally guarantee that this plant is not difficult to propagate. All one need do to establish it (and I am not exaggerating) is accept a bouquet of the dried pods from a friend and walk through the garden. The moonwort will take care of the rest.

Once the moonwort laid claim to my gardens, I didn't have the heart to uproot it. First, the inconsequential but sweet little rosettes of leaves sprang up in bare places that needed the welcome softening of green. I almost forgot the moonwort until the next spring when tall branched stalks appeared flaunting pink and mauve flowers in the midst of my island of snapdragons and ceanothus. I don't know how many days passed before I looked at that patch of moonwort again, but it looked like in the intervening time someone had worked magic on the plants. Flat oval pods, which resembled green silk fairy purses, had replaced the small cruciform flowers. Those spangled pods hooked me. Native or not, the moonwort "passalong" became an accepted and beloved part of my garden.

By autumn my beautiful fairy purses looked pitiful. Brittle beige stalks sported moldy, gray-brown pods that I cut and hung in my herb-drying room. Within a few weeks, papery husks and thousands of the round, black seeds littered the table beneath my moonwort bunches. One rainy afternoon in late November I shut myself inside the workroom and spent hours gently slipping the remaining paper covers off the moonwort. As I worked, I planned for their future.

Lunaria (Latin for moon) annua (for annual) is a half-correct name for this plant. Once the brown paper slipcovers are removed from the dried seeds, their luminous, moonlike appearance speaks for itself, but as for the rest of their name, a taxonomist somewhere made a small mistake. As the saying goes, "The annuals we plant each spring, they perish in the fall; biennials die the second year; perennials not at all." But every Lunaria annua (formerly L. biennis) I've grown, and by now I've grown thousands, flowers and sets seed the second year.

I sometimes think that moonwort has as many legends and names as seeds. In the fourteenth century Geoffrey Chaucer mentioned "lunarie" as a plant used in magical concoctions. In the late 1500s Michael Drayton wrote, "Enchanting lunarie here lies, in sorceries excelling." In the Victorian era Reverend Hilderic Friend ascribed "lunary" with the power to put evil and monsters to flight. He reasoned that since evildoers despise the

• How to Grow Moonwort •

To GROW: This plant thrives in Zones 5 through 9 in dappled shade to sunshine. Sow the seeds in warm, moist potting soil in containers or direct sow into the garden in the spring after all danger of frost is past. Allow 10 to 12 inches between the plants (they can reach a height of 36 inches). Mulch plants to conserve moisture, but keep mulch off the stems and crown of the plant. Feed every two weeks with a combination kelp-fish emulsion liquid fertilizer. Take care to water the plants with clear water before each feeding to be sure the soil is already moist.

To harvest: Wait until your plant stalks have turned brown before cutting them at the base. Loosely bunch stalks and tie with string or secure with rubber bands. Hang bunches upside down until thoroughly dried. When dry, separate stalks and gently slip the brown, papery covering off the pod. There is no need to use preservatives or sprays on the dried arrangements; they'll last for years.

light "lest their deeds be reproved," lunary, in all its shining glory, is a plant for the bad guys to avoid.

My son Noah called moonwort the money plant and the silver dollar bush when its translucent pods appeared. Reverend Friend wrote that the people of Devonshire called the plant silver shilling, moneywort, and money-in-both-pockets. When you take apart the seedpod or look closely at the flower, you'll understand why. Gardeners in Bovey-Tracey referred to it as silks-and-satins, the silks being the green pods of spring and the satins the white of fall.

Passalong Plants coauthor Steve Bender wrote that honesty is yet another nickname for this plant and admitted, "I'd like to tell you how the nickname honesty came about. But, honestly, I haven't a clue." I have an answer for Steve. A woman from Britain once told me that only honest folk can grow this "plant of light" in their gardens. Laura Martin, author and folklorist extraordinaire, has another answer: "The name honesty was given the plant because one can see the seeds right through the transparent disc that holds them." I like both versions and plan to tell these stories when I lead plant-lore walks.

Recorded uses for moonwort are endless. Many old herbals claim that a brew of this plant could pull the shoes off a horse, giving birth to the name unshoe-the-horse. Others give recipes for wound dressings, sallets (salads), and brews to open locks and break chains. Probably the most obscure use cited is in Laura Martin's book Garden Flower Folklore. It states, "The sharp point at the end of the flower was used to prick out notes on thin paper before music was printed," thus earning it the name "prick-song."

Moonwort is as irresistible to children as sunflowers are to goldfinches. When kids discover the ripened pods in the garden, they immediately know they've struck it rich. They hang them from their ears as earrings, make stickpins and brooches, fashion crowns, and gather piles of the gleaming "coins." I love to use the moonwort as fairy dishes and platters for children's pretend tea parties. I spread moss and fern fronds on a stump or rock and fill the dishes with thimble-sized servings of "salad" (flower petals, lichen, and leaves). The main course is "scallops" (hollyhock seeds), and dessert is a "lemon pie" (the center of a daisy).

Throughout the seasons I count on the dependable moonwort to enhance my bouquets, potpourri mixtures (I separate the pods and use them singly), and arrangements of fruits and vegetables. During the holidays I weave the stalks of moonwort into my cedar, redwood, pine, and pepperberry garlands, and poke small clusters of them into the boughs of our Douglas fir and incense cedar Christmas trees. Tiny, clear lights twinkling among the branches accentuate the ephemeral beauty of the translucent pods and make them look like hundreds of glowing white lanterns.

Both my life and my native gardens have changed irretrievably thanks to the kindness of Louise Squibb. Now, every time I look into my patch of snowberries and see the shimmering bangles of moonwort, I think of her and all the good that has sprung from a few seeds.

Felder Rushing and Steve Bender said it best: "You don't need a Ph.D., horticultural library, or yardman to belong to the Passalong Club. All that's required is a piece of earth and a generous heart."

The Bumblebee Rumba

"Once there was a
bumblebee
Who slept 'til spring
had come.'
When the winter broke,
she then awoke,
And her wings began to hum."
—EDITH PATCH, 1926

When I hear the drowsy humming of the season's first bumblebees, I know that spring has finally arrived on our wind-scoured Maine island. This sunny, warm day in May has awakened the queen bees, who passed the long winter hibernating in snug chambers below the soil. I step carefully through the thickets of blueberries to avoid the fuzzy browsers sipping at the blossoms.

Though I'm perched on a nub of rock six feet from the blueberries, I can clearly hear the sonorous buzzing and humming that earned the plump bees some of their common names. "Bumblebee" comes from the Middle English word *bumblen* (to hum). In Germany the bees are called "hummels," and in Britain the allusion to their humming is reflected in the name "humblebee." Bumblebee, humblebee, whatever you choose to call our fifty species of North American bumbles (*Bombus*), they're some of the best pollinators in Mother Nature's workforce.

Two years ago my husband and I made a commitment to the bumblebees who frequent our Maine island. We vowed to stop beheading the dandelions carpeting our driveway and to allow the jewelweed, asters, thistle, and goldenrod to reclaim their territory in the meadow. Along the edge of our narrow oceanfront porch, we planted dozens of pots of vegetables, herbs, and flowers sure to please the bees. We wanted to provide a procession, from spring to the first frost, of nectar- and pollen-rich plant fare. We had an ulterior motive, too—we love tomatoes, and bumblebees are an insurance policy for an abundant, healthy crop.

That spring, our arrival on the island was heralded by some of the coldest, stormiest weather in the past 100 years. We gardened in short spurts between rainstorms, and the fierce winds off the Atlantic laid the flowers and spindly tomatoes down like new-mown hay. Despite the cold and rain the bumblebees were always out and about in the garden, defying the old adage that says "When the bees homeward fly, wind and rain are nigh."

Introduced honeybees are not able to fly at temperatures below 60°F. A prolonged cold snap could prove disastrous for agriculturists and gardeners who depend on the honeybees to pollinate their crops. But our fur-clad native bumblebees, with wing beats of nearly 200 strokes per second, have a metabolic rate about double that of hummingbirds. They are mini flying furnaces, who abide by the postal motto "Neither rain, nor sleet, nor gloom of night...," foraging through cold and into darkness, finding their way home by utilizing the polarized light of the sky.

Bumblebees are great opportunists. They cruise fields and meadows until they find a piece of real estate that suits them, then move in and set up housekeeping. The real estate they favor is usually the abandoned nest of a chipmunk, mouse, or vole. The bumbles rearrange the rodent's abandoned home, tugging at wayward grasses until they have a cozy nest with thick, insulating walls. Some species of bumblebees are ground nesters who build their residences of available plant fibers. Others choose holes in trees, crevices in rock walls, woodpiles, or man-made wooden nesting boxes.

The idea of encouraging the bees to nest on our property intrigued me, so I called Brian Griffin, who wrote the book Humblebee Bumblebee. Brian builds and sells a wooden, two-chamber bumble nesting box with a removable lid. A deluxe observation model is also available—with a "bee tunnel," a booklet, and an acrylic sublid that allows you to eavesdrop on your bees. I'm usually a fairly sensible shopper, but when it comes to critter housing, I want it all. Before I hung up the phone, I'd ordered everything.

The package from Brian arrived with a brief, newsy letter that urged me to "Locate a mouse nest and tuck it inside the chamber. This will almost guarantee success." The search began. Where are the mouse nests when you really need them? In my California barn, I trip over them every time I rearrange tools, but here in Maine, I

• The Buzz on Attracting Bumblebees •

Plant a diverse array of flowers, vegetables, and herbs with blooming times from spring to first frost.

Plant native grasses and wildflowers, and allow a portion of your garden to be a "wild spot" for foraging and nesting bumblebees.

To enjoy the bumblebees' energetic rumba, plant tomatoes, cranberries, blueberries, kiwis, and eggplants.

Avoid pesticides, herbicides, and fungicides; if you MUST spray, use a soap-based, environmentally friendly product and apply it at night, not when the bees are foraging.

Construct a rock pile or woodpile in a shady, seldom-used area of your garden. Do not move rocks or wood once the pile is established.

In a sheltered, shady area, set a flat rock on the ground. Lay a handful of cotton or fine, dry grasses on the rock. Set a 6-inch pot over the grasses; or drill a $5/8$-inch entry hole (large enough for bees, but too small for mice) in the side with a masonry bit. Set a brick or flat stone on top of the pot to cover the drainage hole (bee nests MUST be kept dry) and to discourage marauding skunks.

searched the woods and fields for two days without any luck. In desperation I scraped clean lint from our clothes dryer, pulled old cotton batting from a porch pillow, and stuffed them inside the box.

My husband and I carried the box outdoors and, following Brian's directions, settled it into a low shelf of rock on the sheltered, northern side of our cottage. We were optimistic.

Every day I watched as the queens rumbled along the tops of the swaying grasses in search of a nest. Our bee population was flourishing in the meadow, and the porch garden was abuzz with activity. The bumblebees peeked into the spout of my teakettle-turned-watering can, investigated a hole in the porch rail, and buzzed in and out of our woodpile, but they never approached our bee home. Whenever I lifted the wooden lid of the nest box to peer through the clear observation ceiling, I became the observed rather than the observer. A large brown spider sat straddle-legged across the cotton and lint, guarding her luxurious new territory.

Although our great nesting experiment seemed to be failing, I felt victorious every morning when I visited our pots of tomatoes. The year before, our tomato crop had consisted of five misshapen fruits—tiny, mottled, and as tasteless as pieces of damp cardboard. But after the meadow restoration and the planting of our porch garden, things changed dramatically.

Tomatoes produce their best crops when they are sonicated, or vibrated, by bumblebees. Sonication is a fancy name for what I call the "bumblebee rumba," a rough-and-tumble dance that will induce the most hesitant of tomatoes to set fruit.

I love to sit quietly by my pots of tomatoes and watch the bees in action. They zoom unerringly to a new star-burst bloom loaded with pollen, and grasp it tightly with their jaws and legs. The bumble snuggles against the anthers and begins turning, shivering, vibrating, and buzzing. The wild, twirling dance is so powerful that she must bite into the anthers or else she will shake herself off the flower.

Inside the anthers, the pollen is agitated until it erupts from the pores and settles on the bumblebee's fur. The pollen-covered bee grooms herself, catlike, rubbing her legs across her furry body and pushing the pollen down to a smooth spot on her bristly rear legs.

Workers and queen bumblebees sport a shiny, hairless area called a corbicula (pollen basket). This smooth spot on the outside of a bee's leg is surrounded by inward-curving hairs that hold a cargo of pollen in place. The bees sometimes pack their corbiculae with a load equal to 20 percent of their body weight—that's a lot of groceries in one basket. The bumbles fly to their nest with the pollen, unload the provisions, and head straight back to the tomatoes for an encore. I never tire of watching their energetic performances.

This spring we arrived in Maine with a car full of books, clothes, and one box cradling the remains of a tidy little mouse nest. The humble bumble home is being repaired and set out on the ledge, and we're adding some other pollinator homes we made in our workshop.

We are still optimistic about attracting bumblebees into our boxes, but if we fail again this year, it won't matter. Our meadow is prospering, and the driveway looks like someone sewed golden buttons over its green ribbon of grass. Near the blueberries, a line of empty pots awaits fresh soil and seedlings. Before long the tomatoes will issue their starry, yellow invitations to the bumblebees, and the magical, productive rumba will resume.

"Bee Prepared"

"The more things thou learnest to know and enjoy,
The more complete and full will be for
thee the delight of Living."

—PHALEN

L ast year, on a stormy midwinter afternoon, I sat surrounded by a mountain of unanswered correspondence
and unread books, and slowly leafed through an enticing garden catalog filled with luscious full-color photos and lyrical
retail prose.

I was intrigued by a photograph of something that looked like a birdhouse. It had not just one entry hole, but row upon row of small ones. The caption under the photo stated, "Just hang this bee apartment in your garden and watch as beneficial native blue orchard bees (Osmia lignaria), also known as orchard mason bees, take up residence."

I love bees, whether they are the imported honeybees (first introduced to our country in the early 1600s), fuzzy-rumped bumblebees, or native bees. Since it is illegal to keep a man-made hive in my downtown garden, I felt the houses would be a good substitute and ordered two of them.

A few weeks later my bee houses—two tall, copper-roofed blocks of wood—arrived. I mounted them in a sunny spot on a cedar fence enclosing the children's garden. During the next few weeks, I lurked, watched, and waited patiently for something—anything—to happen in my bee houses. I became alarmed by the absence not only of native blue orchard bees, but also honeybees and bumblebees. Reading the newspaper accelerated my worries when an article cited such a shortage of pollinators that farmers were forced to rent bee hives from out of state.

My lack of bees wasn't an isolated incident, but something being felt by other gardeners and farmers throughout the country. Gardeners know that without the help of a big workforce of insects we won't have apples, pears, almonds, cherries, seed-filled sunflowers, plump squash, and innumerable other crops. As research entomologist Dr. Stephen Buchmann recently stated in a *Tucson Citizen* article, "For every third bite of food we ingest, we can thank our wild-insect pollinators."

I left the little bee houses hanging on the fence and set out to learn about the tenants who never arrived. I turned to Dr. Philip

Torchio, renowned research leader of the USDA-ARS Bee Biology Lab at Utah State University in Logan, Utah, to answer my questions about the tiny bees he has been studying since 1970.

I told Dr. Torchio that I wanted the native blue orchard bees as pollinators and as an educational tool in my children's garden. Enthusiastically he said, "These bees are magnificent for teaching kids about nature in their own backyards. Trap nests are easy to make and children can have exciting personal interactions with the bees." When I told him that most of my visitors are afraid of being stung, he said, "I have worked with these [blue orchard] bees for twenty-five years and have never been stung. Pressure could make one sting, for instance, if it became trapped in clothing, but other than that they are benign."

Dr. Torchio said that blue orchard bees don't live in hives like the honeybees, aren't territorial or aggressive, and can work almost anywhere in the lower forty-eight states, except in areas of climatic extremes such as low desert or high altitudes. He began recommending these overlooked bees as effective alternative pollinators long before the current alarm over the loss of honeybee colonies. Since the 1940s, those colonies have been reduced by 50 percent and their numbers continue to dwindle as Varroa and tracheal mites take a further toll.

Dr. Torchio cautioned, "Even the blue orchard bees are not doing well in our urban environment. We have a shocking lack of green belts and an incredible amount of chemicals being sprayed on lawns and gardens. These bees are our canaries in the coal mines." Miners knew to make a hasty getaway the minute their gas-sensitive canaries fell from their perches, but our blue orchard bees don't give such a dramatic warning. Instead, the small workers disappear silently, almost imperceptibly, missed only by scientists, gardeners with empty bee houses, and farmers with barren fruit trees and seedless sunflowers.

On chilly early-spring mornings, when the imported honeybees snuggle inside their hives, our hardy, short-lived, native blue orchard bees are out and about in orchards and gardens as soon as the temperature reaches 55°F. Not only are they hard at work before the honeybees clock in in late spring, but they are more efficient and effective pollinators. Honeybees keep their pollen cargo neatly contained and tucked into small baskets on their legs. The blue orchard bees, virtually bristling with pollen that has collected in the hair on their abdomens, scatter their payload as they visit each bloom. The honeybee tribe locates one food source and stays with it throughout the day; this itinerant blue orchard bee ranges far and wide—the better to facilitate cross pollination.

As Dr. Torchio and I talked, I sadly recounted my failed bee experiment at the Heart's Ease Gardens. He modestly told me that he has drilled and tested more than 2,000 wooden nesting blocks to determine what size and configuration of holes works for the finicky bees. He explained that the blue orchard bees do not excavate their own holes; instead, they cruise the landscape looking for an opening that is just the right size (about $5/16$ths of an inch) to accommodate their bodies, then use their famed masonry skills.

My copy of Insect Architecture (1830) describes how female bees collected mud, clay, and bits of chalk from a cart rut in a road to construct the walls of their cell. Dr. Torchio explains that "the females then collect pollen and nectar and pack it in the cell until it is about half-filled. An elongate egg is then deposited on the surface of these pollen-nectar provisions just before the cell cap is constructed."

After the egg is laid and the cell is partitioned, or "capped," with moist soil, the industrious bee continues constructing more cells and laying more eggs until the hole is filled and plugged with a thick, protective wall of mud. Entomologists have discovered that the first eggs laid in the back of the cell are the female ones. Perhaps the females are positioned the farthest back to better protect them from the predations of other insects and the probing beaks of hungry flickers and woodpeckers.

"Constructing traps of wood gets to be an expensive and time-consuming project," Dr. Torchio said. Trying to drill, store, and clean hundreds of nesting blocks is not economically feasible for a grower with thousands of trees. To make blue orchard bees a viable choice for commercial growers, Dr. Torchio has spent the past five years experimenting with paper nesting traps. Bud Miller of Custom Paper Tubes in Cleveland has worked with Dr. Torchio in developing a hygienic, lightweight, affordable trap nest manufactured from recycled fibers. Each five- by eight-inch cylindrical nest wears a protective coating of Mylar and is filled with fifty-five removable paper-straw inserts that the bees use as nesting chambers.

After the bees have laid their eggs and plugged the open ends of the inserts with mud, the straws may be removed and stored in a protected area. The empty cylinder can then be refilled with new straws. One cylinder properly restocked could cause a quantum leap in the blue orchard bee population in your garden. Wooden traps can become infected with parasites and are subject to predation by insects. The new paper traps with removable inserts reduce that threat.

How to Please a Population of · Native Blue Orchard Bees

In the early spring, hang the clean nesting blocks and the tubes of straws, with their entrances facing the morning sun, in a protected area.

Set aside a "safe zone" underneath the houses so that the emerging females who fall onto the ground aren't trampled by curious or oblivious passersby.

Make sure to have a taste-tempting floral feast that is totally free of any herbicides, fungicides, and pesticides (that is always a rule in my gardens).

Place a shallow terra-cotta saucer, mounded with moist soil, a bit closer to the bees' home so that they have plenty of material for their construction projects.

Never take the miracle of pollination for granted.

One or two weeks before the projected bloom time in an orchard, the straws can be re-inserted into the cylinder and mounted in a sunny, protected area near the crops requiring pollination. This deceptively simple trap may be the tool that will enable orchardists to ensure both abundant crops and future generations of the beneficial bees.

When I bite into a voluptuous pear, or nibble on a crunchy almond, I'll say a thank you to dedicated scientists such as Dr. Philip Torchio, and to the tiny, invaluable blue orchard bee.

Flower Flies
Forever

Take the time to know these
tiny beneficial insects.

How do I forget from year to year the unexpected beauty and life that bursts from the garden each spring and summer? I can't take a step without discovering a newly awakened plant or the arrival of winged visitors. Every morning I look forward to my walk along the pathways, and I force myself to forget the chores and focus on the miracles unfolding around me. I try to remember the words of artist Georgia O'Keeffe, who once wrote that people don't see things because "to see takes time, like having a friend takes time."

I spent more than an hour early this sunny day watching the busy air traffic around my terra-cotta pots of coral bells (*Heuchera* spp.) and mixed wildflowers and herbs. I trailed behind a few of the tiny black-and-yellow-striped flower flies, also known as hover or syrphid flies, which hang in the air above blossoms as though dangling from an invisible thread. These welcome visitors are responsible for both pollinating and protecting many of the plants in the garden.

North America is home to more than 900 species of flower flies, but members of this huge family are found in many parts of the world. If you have an array of flowers in your garden or an outbreak of aphids, there is a good chance flower flies are already a part of your landscape. These members of the order Diptera (from the Greek for "two-winged") resemble the four-winged bees and wasps and carry out some of the same functions. But they live a solitary lifestyle, rather than in an organized community with divisions of labor and castes, and they are stingless and docile enough to hold in your hand.

Every time I crouched near the coral bells, the big-eyed flower flies backed out of the blooms and sidled away to a new plant. I was frustrated and decided to use a tactic taught to me by Audubon Society naturalist Frank Gander, who said, "Sometimes you just need to sit quietly and wait for nature to come to you." I moved the container of coral bells next to a sunny bed of feverfew and coreopsis and stretched out on the ground beside them. As an afterthought, I grabbed a small pot of foamflower (*Tiarella*) and white bacopa and set it on my chest.

· Flowery Food for the Flies ·

Adult flower flies need not only nectar, but also plants rich in pollen, which is a source of the fats, vitamins, minerals, and proteins essential for reproduction.

Cultivate flower-fly–friendly plants

With mouths that are shorter than those of bees, these pollinators seek out small flowers and open-faced blooms. Plant some sweet alyssum, anise, asters, bacopa, cilantro, coral bells, coreopsis, daisies, fennel, feverfew, lobelia, parsley (let it flower), small-flowered sunflowers, foam-flowers, and veronicas. Even a hefty container or window box, which also makes viewing easier, provides enough food for dozens of flower flies.

Man-made mud puddles

Fill a shallow glazed container or saucer with a mixture of sand and mud. Place it under a dripping faucet in a sunny area, or moisten the mixture daily. Pollinators of all types will stop to sip moisture and minerals from the soupy soil.

Pollinator lovers beware

Avoid the use of pesticides, herbicides, and fungicides. If you spray aphid colonies with homemade remedies or water, always check to make sure you are not injuring beneficial aphid tigers, which resemble small, translucent, yellowish green maggots. These little insect vacuums develop in aphid colonies and consume hundreds of aphids daily.

Within minutes, one of the little ($1/4$-inch to $1/2$-inch) pollen pirates approached, and I lifted my head to watch it alight on a bloom. It paused to "taste" the nectar with the receptors on its legs, then lowered its short, tubular proboscis into the flower. After a moment of sipping, it slid backward, and I could see a faint dusting of sticky pollen on its hairy chest. It flew closer and hovered above me like a helicopter, then darted up a ladder of blooms and into a new flower. As it moved among the blossoms, it traversed an ancient floral "trade route," transferring pollen from the stamens of one flower to the stigma of another—and thus ensuring future generations of both flower and fly.

Where there is abundant, healthy food (flowers and aphids) and no pesticides, male and female flower flies will engage in the business of courtship and mating. After mating, the female does a reconnaissance flight through a garden in search of colonies of aphids. When she locates an active infestation, she dips her abdomen into the midst of them to oviposit one of her elliptical, white eggs ($4/100$ of an inch), then sets off in search of another likely site.

Flower flies go through a complete metamorphosis in two to four weeks, sometimes producing as many as five to seven generations a season. During the larval stage, the blind, translucent, yellowish-green aphid tigers, which resemble maggots, don't stray far from their homesites. Instead, they hang onto a leaf with their hindquarters and, as described by scientist B. J. D. Meeuse in his fascinating book The Story of Pollination, sway back and forth "somewhat like the trunks of elephants" until they come into contact with an aphid, mealybug, caterpillar, thrip, scale, or leafhopper. As soon as an aphid tiger brushes against prey, it grabs hold, pierces the victim with its mouth hooks, and sucks out its body fluids. An empty carcass signals the end of their meal and triggers their deceptively haphazard, swaying "dance" for food to begin again. A single aphid tiger can devour hundreds of garden pests during its development and is rivaled in effectiveness only by the larvae of ladybird beetles and lacewings.

Effectiveness isn't the only attribute of the generations of aphid tigers that inhabit a garden; they also have longevity. Before the first larvae of the lacewings and ladybird beetles emerge in the spring, the aphid tigers are already "dinner dancing" through the garden. When temperatures cool and other predators disappear in the autumn, the last generation of aphid tigers still will be there to fight the good fight, feeding on unwanted pests until nature's time clock and thermometer hasten their overwinter pupation or rest.

As the sun rose higher, my warm patch of light became shaded, and the flower flies moved on to richer, warmer feeding grounds. I lifted the small pot off my chest, brushed some soil from my nightgown, and slowly arose. Perhaps I lost a morning of work, but I finally got a chance to see a flower fly from the perspective of one of its own. I now feel as though I finally understand a little more about this garden friend. That took not only time, but also a measure of patience—two essential ingredients in the world of a garden.

How to Enchant a Wasp

*"Do you have anything kind to say about wasps?
They flourish on our patio, and they don't attack
us or the pets. They remind me of faeries,
and I love the way they hang in the air."*
—A LETTER FROM DOROTHY CLENDENIN

Meals on our small seaside porch are always an adventure. As soon as we sit in our rockers, the jays, red squirrels, wasps, and yellow jackets stop for a handout. They feel entitled to a share of everything we eat outdoors. To keep them satisfied and otherwise occupied, we set a plate of sunflower seeds on the far end of the porch railing for the birds and squirrels, and a tiny slice of ham and a dollop of Maine blueberry jam for the yellow jackets and wasps.

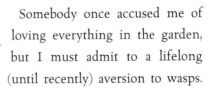

Somebody once accused me of loving everything in the garden, but I must admit to a lifelong (until recently) aversion to wasps. They were always enigmas to me and the bane of my summers.

My early acquaintance with paper wasps (*Polistes*) was intimate and painful. Every morning Mother sent me into the yard to collect the fallen fruits under our apricot trees. The sweet juice of ripened fruit is as attractive to wasps as nectar is to hummingbirds, and large numbers of these insects congregated under the trees early each day.

The graceful wasps, members of the order Hymenoptera (membrane-winged), cruised peacefully above the glowing apricots nestled in the grass, then sat atop them and lapped their juices like kittens at a milk bowl.

I soon learned, after grabbing an occupied fruit, that wasps can and will sting repeatedly if they are annoyed. Unlike the barbed stinger of a worker honeybee, which remains in a wound, the wasp's stinger is finely barbed and can slide in and out effortlessly.

For two spring seasons of my childhood, my best friend Ricky and I concentrated on eating every green apricot we could find. I thought that if we stripped the trees before the fruit ripened, my daily conflict with the wasps would end. No matter how thoroughly we searched, hidden fruits eluded us. They plumped, colored, ripened, and fell, and my yearly battles with the wasps continued.

Scientists now know that social wasps use both landmarks and odors as guides to return to their most successful foraging sites. Monica Raveret-Richter, from the Department of Biology at Skidmore College in New York, wrote that experience and the presence of other wasps influence the feeding behavior of individual wasps. Scientists call this "social facilitation," but I think of it as just following the crowd. Hours after I removed every fruit from under the apricot trees, the wasps returned and glided over the grass in search of their misplaced meals.

Nowadays, the wasps and I enjoy a peaceful coexistence. I look forward to spring when the mated young queens, who overwinter in leaf litter or sheltered crevices, emerge from their hideouts and begin the task of nest building high in our protected north-facing eaves.

The wasps sail past our porch and stop to tear tiny pieces of weathered wood from a post. They use their strong jaws to manducate, or chew, the fiber mixed with their saliva into a papery pulp. Then, the magic begins. They attach a small pedicel to the eaves and carefully construct what Cornell University professor Anna Botsford Comstock termed "their gray apartment house." In her

• Emergency First Aid •

- Apply ice and lie down.
- Watch for symptoms such as swelling, hives, wheezing, or faintness, which may indicate an anaphylactic shock reaction.
- Seek medical aid if necessary.

• Be Wasp Wise •

Paper wasps are considered docile, but they can be a danger to humans if they build their nests near doorways or frequently traveled pathways. In these cases, it is wise to remove the nests as they are being built (when the wasps are out of the vicinity). You may have to do this more than once, but they'll get the hint and move elsewhere.

Wasps sense when poisons are being applied to their nests and will attack, sometimes even at night.

Never swat a wasp. They secrete a chemical that signals danger to other members of their colony, who will attack in defense of their nests.

If you find yourself in an area with high wasp traffic traveling in an established pattern (beeline), move on slowly; you are probably near a wasps' nest.

Keep soft-drink cans covered or wasps may enter the drinking hole.

Don't wear perfume.

Don't go barefoot.

Believe it or not, brush your teeth if you've eaten strong-smelling foods such as tuna, and make sure to wash around your mouth and hands, so as not to attract the unwanted attention of wasps.

1903 book *Ways of the Six-Footed,* Comstock credits the wasps, who vie with the honeybees for the title of nature's finest architects, as being the first papermakers.

When wasp traffic on our porch increases, we know that their white grublike larvae fill the paper-celled nests. A demanding feeding schedule keeps the wasps busy throughout the day. They glide into our container gardens and seize some of the worst of our plant pests, skin or chew the critters, then feed them to the young grubs.

Bill and Helga Olkowski and Sheila Daar, who wrote the comprehensive and invaluable *The Gardener's Guide to Common-Sense Pest Control,* extol the virtues of the maligned social paper wasps. They note that researchers are actually building artificial nesting locations to increase the populations of Polistes, and to measure their predation rates on agricultural pests such as cabbage butterflies (*Pieris rapae*) and tobacco hornworm (*Manduca sexta*). The authors cite studies of a new colony where individual queen wasps killed between one and eight cabbage butterflies a day. Those numbers are impressive, but as soon as the wasp workers emerged from their nests, the numbers topped 2,000 caterpillars per colony per day.

Although these smooth-bodied wasps are not usually thought of as pollinators, many species do serve that important function. Coarse hairs on their legs trap tiny pollen grains, which are transferred between plants as the wasps sip at nectar-filled flowers.

Autumn is approaching and the wasps are still paying us visits. We regard each other warily, though I still seem to be the most skittish in these contacts. Although I was stung many years ago, I still remember the searing pain and how the wasps chased me through the gardens and into my grandmother's back porch.

I respect these hardworking wasps and appreciate the job they do in the garden, but I've drawn the line as to how we conduct our relationship. Wm. Hamilton Gibson, author of the nature classic *Sharp Eyes* (1891), devoted an entire chapter on how to handle a wasp. Gibson claims that September is the best time to hold them and that there is a certain knack involved in handling them that can be mastered by anyone. Creep up slowly, he advises, hold your palm open and murmur this enchantment, "*Polistes! Polistes! Bifrons! Proponito faciem!*" Gibson cautions you to wait until the wasp turns toward you, "then with a quick clutch grasp your prize." He claims that "no amount of abuse will induce the wasp to sting," and that "perfect faith in the charm will enable any one to handle a wasp with impunity." I think I will continue to admire the wasps from afar, and just make sure that their autumn supply of blueberry jam is kept well stocked.

Beetles to the Rescue

*There's nothing like a working flock of ladybirds
to keep a garden healthy.*

E arly each summer morning at our cottage in Maine, as the sun peeps above Pemaquid Point and spills its rose-tinged light across the waters of John's Bay, I carry my mug of coffee outside and begin my day. A gentle breeze stirs a line of old wooden porch rockers, which tip slowly back and forth in silent greeting. Their wide, woven seats are tempting invitations to sit and linger, but before I settle down I always wander among my plants and give them the first of their daily inspections.

Drifts of pale blue trailing lobelia, sweet alyssum, and white bacopa billow like multicolored petticoats from a half-dozen long terra-cotta troughs. Sunflowers, zinnias, anise hyssop, nicotiana, cosmos, and herbs mingle intimately in the pots, vying for their share of the sunshine. I know that when plants are crowded, disease and pest infestations are more likely to become a problem—but if I remain vigilant, the garden flourishes.

For the past few months, the porch-grown rainbow of flowers and herbs has supplied us with dozens of colorful bouquets and flavorful meals. Last week while I traveled, as though they knew that I wasn't there to protect my plants, the aphids arrived en masse. When I returned home, I found my basil and tidy line of lettuces as flat as punctured balloons and crusted with aphids.

Normally I don an old pair of gloves and run my hands over any aphid-infested plants to dislodge the pests, but as I looked closely I could see that the aphid situation was now under control. I watched as a shiny red ladybird beetle grazed on a succulent mass of celery-green aphids. While many people call them "ladybugs," they aren't actually bugs; they're beetles with differing mouth parts and wings.

North America hosts about 400 species of these beloved beetles, which are known by such colorful names as cush-cow lady, Our Lady's ear, Barnaby bug, red-turtle-beetle, little-spotted-elfin-cow, and spotted-tortoise-of-the-dell. Whatever you choose to call them, these members of the Coccinellidae family and their fanciful alligator-like larvae are small blessings for every gardener.

Each ladybird beetle consumes hundreds of aphids before settling down for some serious reproduction. Throughout spring and summer, you may watch as they deposit their millimeter-long, yellowish-orange, football-shaped eggs on the undersides of leaves.

Within three to five warm days, the alligator larvae, called aphid wolves, hatch and begin their moveable feast. After devouring a few hundred aphids and going through three molts, the larvae attach themselves to leaves and continue their metamorphosis as they enter the pupal stage, which lasts about a week.

One warm and foggy June afternoon, as I worked my way down the plant-lined porch with my watering can I discovered a ladybird beetle emerging from its pupal skin. To celebrate its birthday, I picked a pitiful white nicotiana bloom encrusted with aphids and set it beside her (and I use that appellation loosely since I couldn't identify the sex). Within a couple of minutes, she climbed aboard my offering and began to gobble aphids as quickly as a kid sampling free candies in a chocolate shop.

When I was a child in my grandmother's garden, the ladybirds were one of the joys of my life. I spent countless hours playing with them and their larvae, which I called "Halloween bugs" because of their black-and-orange coloration. Many years later, remembering those good times, I decided to buy my son Noah a carton of ladybirds to release into the garden for his fourth birthday. Noah opened the flaps of the small gift box, and dozens of ladybirds crawled out of their bed of excelsior and up the sleeves of his denim shirt. His big, cinnamon-brown eyes widened, and he laughed gleefully. "This is my best birthday surprise ever," he said, as he lifted out another glistening handful

· Natural Pest Control ·

Ladybird beetles, which feed on aphids, scale, mealybugs, white flies, and mites, can be one of a gardener's best friends. Entice them into your yard with a diverse array of pollen- and nectar-producing plants such as tansy, fennel, dill, angelica, clover, coriander, yarrow, sweet alyssum, chervil, and sunflowers.

Don't panic if you find large numbers of yellow eggs attached to the undersides of leaves; they will be the future generations of ladybirds. Watch closely for three to five days, and you may see the emerging alligator-like larvae. These small predators earned the nickname "aphid wolf" because of their voracious appetites.

Encourage the presence of beneficial beetles and their offspring by refraining from the use of any pesticides. Give these beetles (and their kin) a chance and a poison-free environment, and they'll work for you.

Before "washing" aphids off plants, always check to make sure you are not disturbing ladybird beetles or their larvae. If you find them feeding on aphids, and you can't stand to have the infested pieces of plant in your garden, simply snip the offensive portion off the plant (making sure you don't dislodge the beneficials) and hide the clipping under a bush.

> ● A Recipe for Sweet Temptation ●
>
> In field tests conducted by Utah State University, a sugar water mixture was sprayed around the edges of alfalfa fields. Researchers found that this encouraged visits by ladybird beetles, and many other beneficial species as well.
>
> 1. Stir 5 ounces of sugar into 1 quart of water.
> 2. Shake well and transfer the syrup to a spray bottle.
> 3. Spray syrup onto aphid- or scale-infested plants daily.
>
> **Caution:** Take care never to spray this sticky mixture directly onto any beetles, as it will cement their wings closed.
>
> **Note:** Peaceful Valley Farm Supply, Inc. sells a product called Good Bug Food, which, when set among crops, attracts many beneficials.

of the beetles and released them into a fragrant patch of blooming bee balm.

Since that day, I can't look at a ladybird without thinking of my son and his heartfelt response to a simple gift. This year, for his twenty-ninth birthday, and I suppose for sentimental reasons, I plan to send him a carton of aphid wolves (the Halloween-colored larvae) from a mail-order beneficial-insect supplier. Each order comes with instructions, equipment, and food to rear the larvae to their glossy adulthood. Perhaps, as he releases his winged progeny into his shoebox-size city garden, he will remember that birthday of so many years past.

• How to Herd Your Ladybirds •

If you order a supply of ladybirds for pest patrol in your garden, take these simple steps to ensure success.

1. Purchase your ladybirds from a reputable dealer who either rears them in an insectary or screens the field-collected beetles for parasites (important because infected beetles can spread disease to a healthy garden population).

2. Check with your supplier to make sure that you will not be receiving hibernating beetles, who must use their stored body fat by flying before they begin to feed. They may end up in your neighbor's garden instead of yours.

3. Request that suppliers ship your order via Priority Mail or overnight; and ask them to ship only on Monday or Tuesday, so the beetles don't have to endure a long weekend in a box.

4. Water your garden liberally (overhead sprinkling works well) before releasing the thirsty ladybirds.

5. Release your ladybirds in the evening or at night. This will discourage them from abandoning you the moment they are freed.

6. Don't place all your ladybirds in one area, but do locate them near aphid-infected plants. Typically, only about half a dozen ladybirds live in 10 square feet of garden. They are voracious eaters and need plenty of room to roam and graze.

Dessert and a Dragonfly

"Deep in the sun-searched growths
The dragon-fly hangs
Like a blue thread
Loosed from the sky."
—CHRISTINA ROSSETTI

The brigade of dragonflies patrolling the old Cove Road has complicated my life. A trip into town that should take less than twenty minutes now takes forty since I can't drive without stopping every few hundred feet to observe the jewel-toned dragonflies skimming above the road and pond, "hawking" insects from the warm blue skies.

My fascination with these ancient Odonata (their fossil remains date back 300 million years) began in my Grandmother Lovejoy's sunny California garden. Her bountiful perennial beds were planted with crayon-colored hollyhocks, purple iris, columbines, and a carpet of dianthus and sweet alyssum. Dragonflies kept vigil over the insect-busy beds, constantly patrolling and engaging in wing-clashing territorial disputes with each other.

As a child I loved to climb my favorite sycamore tree in her garden, drape myself over a smooth, swooping limb, and peer down on the frenetic activities of our resident dragonflies. Garnet, green, orange, and blue-green species zigged erratically over the pathways, some joined together tandem style, like interlocking pieces of a jigsaw puzzle.

Grandmother told me that dragonflies are called "Darning Needles" for their long, narrow bodies. She said that people once believed that these sun-loving insects disappeared on cloudy days because they were busy embroidering rainbows onto the framework of the sky. Each brilliant band of the arch represented the color of the "darner" that had dutifully done its job. Even today I can't look at a rainbow without visualizing a quilting bee of colorful dragonflies stitching industriously.

In the heat of a recent late-summer day, I turned off the highway onto Cove Road to gather a basket of berries. I parked my car, grabbed the clippers, and began a hike along a roadside humming with insects and glistening with the darting maneuvers of the dragonflies. I found a patch thick with berries and quickly filled my harvest basket. As I turned back toward my parking spot, a car sped past and something landed at my feet—a spotted calico pennant (Celithemis elisa) lay stunned but intact.

Although I had spent hours crouched stoically alongside ponds and streams and had observed dragonflies through binoculars and camera lenses, I had never been so close to a living specimen. I set my basket next to it, knelt down, and looked closely as it moved slightly. A car approached and I waved it on as I picked up the calico and set it atop the plump berries.

Jeff was up on a ladder when I arrived home. "What's in the basket?" he asked. "Dessert and a dragonfly," I answered. He scrambled down and looked at our visitor. "What are you going to do with it?" he questioned. "Nurse it, watch it, and draw it," I said, as I nestled it on a bouquet on our front porch. I moved my worktable, paints, and a magnifying glass into position, set the pitcher of flowers on the table, and sat down.

As a mosquito settled on my arm, I swatted it and placed its remains in front of the dragonfly, who ignored it. Another mosquito landed and this time I whacked it lightly and placed it, still kicking, in front of the dragonfly. It disappeared. Contrary to my preference for meals without movement, the dragonfly wanted its food fresh.

I held the magnifying glass above the dragonfly and used tweezers to place more squirming mosquitoes on top of a scabiosa blossom near its head. Even with the aid of the magnifier, I couldn't see the swift capture and consumption of the prey, but the snacks disappeared without a trace.

During that afternoon I served up more than four dozen mosquitoes, but that didn't set any records: According to an article by Richard Conniff in the July 1996 *Smithsonian Magazine*, a dragonfly can locate prey, zoom in for an attack, and return to its perch for a feast in just over a second. And they do that 300 to 400 times a day.

In between feedings, I used the lens to study my guest. The cliché "he was all eyes" is apropos when describing these amazing creatures. In her exquisitely illustrated book *Dragonfly Beetle Butterfly Bee*, Maryjo Koch writes, "With eyes that extend almost completely over the top of its head, the dragonfly has a virtual 360-degree perspective. Its compound eyes have as many as 25,000 facets, so it makes sense that 80 percent of the dragonfly's brain is given over to analyzing what it sees." Able to spot prey (or territorial interlopers) at distances of up to 100 feet, dragonflies not only analyze but quickly respond to any airborne activity in their vicinity. (Try fooling one by tossing tiny wads of thread into the air. You'll be surprised by just how quickly it can maneuver.)

The *National Audubon Society Field Guide to North American Insects and Spiders* states that the dragonfly's four powerful wings (I have driven alongside one cruising at fifty-four miles per hour) are able to move independently for their typical zigging flight pattern, or in concert for rapid-fire responses. If you watch a dragonfly's antics, you will observe it using this aerodynamic ability to "helicopter"—hover, move forward and backward, and lift off its perch in an instant. It is no wonder that it is such a proficient hunter, plucking its prey from the sky (it can carry a cargo twice its weight) and depositing it in the "basket" of legs folded beneath and in front of its head.

I ran my finger across the stiff wings and back of the dragonfly, which had shifted its position on the bouquet. It was hard for me to believe that this formidable-looking flier, one of 5,000 species worldwide with large wings as thin and transparent as a sheet of mica, had spent two to three years of its life underwater. As a voracious naiad, it had killed and consumed everything from bugs and larvae to tadpoles and small fish.

After years of growing and molting, this fully developed naiad had emerged from its watery world, crawled up a reed or onto a twig or rock, split open its skin, and squirmed out of its former covering like a diver wriggling out of a tight-fitting wetsuit. The rest of its life—the short but action-packed two- to three-week dragonfly phase—would be spent in the pursuit of prey (especially mosquitoes), dramatic territorial clashes; the tandem, interlocked flight of reproductive activities, and, in the case of my visitor, recuperation.

As the calico pennant sat waiting for the delivery of its next meal, I read through my portfolio of field notes. One page was devoted to quotes from the 1924 classic *Precious Bane*, by Mary Webb. In it she describes a phenomenon called "the troubling of the waters," when the clear ponds churn with the emerging bodies of the dragonfly naiads.

The Shropshire, England, villagers of Webb's book revered the dragonflies and referred to them as "ether's [adder's] mon" or "ether's nild" for their alleged ability to detect adders (snakes). They believed that the dragonflies would hover and circle wherever snakes hid, warning passersby of their presence.

Another page of notes cited some historical records from the 1860s recounting the migrations of vast swarms of dragonflies from continental Europe to Britain. Their numbers were in the billions and the sky over Antwerp turned from daylight to darkness as they passed overhead. In the margin I had scribbled a note from Richard Conniff's article: "Cape May Bird Observatory in New Jersey has similar sightings and now offers guided dragonfly workshops."

The dragonfly moved slightly and fixed its multifaceted gaze toward the granite ledge and the water beyond. I leafed through

· Attracting Dragonflies ·

Open your heart and share your garden with a few of the 450 species of dragonflies found in North America.

- Insects are a dragonfly's sustenance. To encourage dragonfly visits, avoid the use of pesticides.
- Situate an old bathtub, large crock, half barrel, water trough, or metal tub in a sunny area of your yard. Pre-formed ponds and pond liners are available, if you wish to install a small in-ground pond.
- Plant a selection of water plants such as a water lily or water primrose (*Ludwigia*) and some submerged plants such as *Vallisneria*, *Myriophyllum*, *Elodea*, or *Cabomba*. These will supply oxygen and provide habitat for dragonfly naiads.
- Stick a tall bamboo stake or twig into one of the sub-merged pots. The twig or stake should protrude a few inches above the water surface so dragonflies can use it as a perch or resting spot.
- Allow a little organic debris to gather on the bottom of your pond. This debris will shelter critters who will be a meal for the voracious dragonfly nymphs.

the portfolio and discovered a page of sheet music tucked inside a thick packet of notes. The music and notes were from a lecture by Ernie Siva, an eloquent Native American teacher. The music was simple and repetitive, a soothing canticle written especially for the dragonflies that frequent the shores of Southern California creeks.

During the lecture Ernie explained that "Uska" is the word his people use for dragonfly. "Just as we chant to our babies to be peaceful and have no anxieties, we chant to the Uska to calm them. They are a keen gauge of our feelings; they know whether we will show them kindness. When we sing to them they come down from the sky and land on our hands." I remembered how Ernie then smiled and how lucky I felt to have learned such an enchanting tradition.

I looked at the sheet of music Ernie had given me and memorized the unfamiliar words: "Ush Kana ush Kana oh oh ush Kana ushk, ush Kana ush Kana oh oh ush Kana ushk." I made a hasty swipe at a mosquito, passed the last meal of the day to our hungry guest, and went indoors for my own dinner.

The brilliance of the sun reflecting off the water and into our bedroom awoke me before six A.M. I ran outside to the porch and peered at the bouquet. The dragonfly was gone. I turned toward the ledges and the undulating mounds of Rugosa roses and bayberries. Above them a dragonfly hovered, dipped, and scooted toward the porch. I stood still and watched as it dropped below the roofline, paused near my face, then turned and zipped out of sight.

After breakfast I picked up a basket and told myself I was going out for more berries, but instead of clippers I grabbed Ernie's sheet of "Uska Lullabies" and tucked it into my pocket. As I turned my car onto Cove Road, I slowed and watched as dozens of dragonflies glinted above the shimmering ribbon of asphalt. I parked, slipped out of the car, and quietly closed the door. I felt like the small, expectant child in Grandmother's garden as I turned, began to sing, and walked into the midst of the dragonflies, my hand outstretched before me.

Calling All Caterpillars

"There is a mystique about caterpillars. It is best appreciated by small children and other dreamers but beyond the ken of pragmatists."
—MICHAEL A. GODFREY, *A Closer Look*

S ide-tracked again. This time by the sight of the feathery foliage that had poked its head above an undulating border of scented pelargoniums. "I swear that wasn't there yesterday," I muttered to myself as I walked toward the tall fronds of sweet fennel. As I neared the border, I saw innumerable green caterpillars, banded in black and sprinkled with a confetti of carnelian orange spots. They looked like brilliant enameled jewelry strewn carelessly over the branches.

I crouched down next to them and watched as the larva of the anise swallowtail (*Papilio zelicaon*) feasted nonstop on the soft new foliage. My son Noah once referred to caterpillars as "eating and pooping machines," and that is just what they are, consuming their own weight in food every twenty-four hours.

Entomologist Julian Donahue describes a caterpillar as a "self-stuffing sausage that gorges itself on leaves or flowers, pausing only occasionally to shed its skin when the casing becomes too tight." These "self-stuffing sausages" are the second stage in the four-phase life cycle or metamorphosis (the Greek word for "transformation") of butterflies and moths.

Though butterflies are beloved by gardeners everywhere, the sometimes comical or fierce-looking caterpillars are often looked upon with disdain. Many gardeners do not realize that only a few species of caterpillars cause significant damage to cultivated plants, and greet their visits with a quick spray of insecticide or a heartless squishing. I'll have to add my plea to that of entomologist Donahue, who asked that we stop treating caterpillars as objects of scorn and start showing a little tolerance.

I picked up one of the biggest caterpillars and stroked it gently. Though the word caterpillar is derived from the Old French *catepelose*, or "hairy cat," the hungry muncher I held in my hand felt more like a fat worm in a form-fitting leotard. Others, like the familiar and beloved woolly bear, definitely live up to their name.

• Planning for a Butterfly •
and Caterpillar Friendly Garden

- Plant natives that are local or endemic to your region.
- Plant nectar rich flowers to feed butterflies, such as cosmos, sunflowers, coneflowers, salvias, milkweeds, asters, butterfly bush, verbenas, zinnias, bee balms, catmints, phlox, goldenrod, scabiosa, thyme, sedum, wallflower, or Joe-Pye-weed. Just look around and notice where butterflies are nectaring.
- Plant larval food species to entice egg laying butterflies. Some of my favorites are willow, dogwood, blueberries, clover, Dutchman's pipe, birch, violets, mallow, and grasses.
- Leave an out-of-the-way area of your garden wild and weedy. You'll be surprised by the number of species enticed to such a spot.
- Build a permanent woodpile or tepee of bark covered logs for overwintering species of butterflies.
- Situate shallow saucers of mud on the ground throughout your garden to provide nutrient rich puddling stops for butterflies.
- Place flat rocks in sunny areas to encourage butterflies to bask.
- Let some of your perennials and grasses, where larvae may overwinter, remain untrimmed until spring—a little untidiness can lead to an explosion of butterflies.
- Never use pesticides or herbicides in your garden, and remember, Bt (Bacillus thuringiensis) will indiscriminately kill ANY feeding caterpillar, not just the bad guys.

I always use caution when handling the hairy caterpillars (these usually become moths) that I find in my gardens or woods. Some, like the members of the giant silkworm family, may have hairs that sting or irritate if they are touched. When I want to identify a specimen, I simply slip a piece of stiff paper underneath it, use a magnifying glass and field book, and then gently put the caterpillar back exactly where I found it.

"If you can't fight them, join them," seems apropos when it comes to gardening in harmony with the Lepidoptera (Greek for "scale-winged") family. Scientists estimate that there are between 112,000 and 170,000 known species of Lepidoptera worldwide. Of these, 11,300 call North America home. Although their numbers are nearly incomprehensible, sightings of both butterflies and moths have diminished drastically in the urban landscape, and have dwindled noticeably in the countryside. I wondered what factors had contributed to the decline and called noted entomologist Andy Calderwood of the Santa Barbara Museum of Natural History.

Andy cited the use of pesticides, urban pressure, destruction of habitat, monoculture, introduction of exotic plants, and air and water pollution as just a few of the causes of reduced caterpillar/butterfly/moth populations. It sounded dismal, but he offered some words of encouragement and assured me that he felt each and every gardener and backyard can make a substantial difference to our populations of Lepidoptera.

I told Andy that I was anxious to plant a garden specifically for caterpillars and asked him for some guidelines...not from a gardener's viewpoint, but that of a seasoned entomologist. He said,  "If you really want to have a garden full of caterpillars, you have to plant both nectar flowers and larval-host plants to attract butterflies. Butterflies cruising over your garden will spot those host and nectar plants and fly in to feed and lay eggs. Once the eggs hatch and the caterpillars begin feeding, there will be even more activity."

As we finished our conversation, I asked Andy to share what he felt was the most important thing gardeners could do for the health of their local butterfly/caterpillar populations. "Plant natives," he said without hesitation. "Plant natives that are local or endemic to your region of the country, and by this I mean don't plant a southern native in a northern garden. This is not a trivial matter," he added emphatically. "If the larvae of a butterfly species depend on a certain species of native plant, and that plant disappears, then we have also lost that butterfly. By planting those species found locally, we can make great strides toward mitigating some of the damage we've already done. And, we'll strengthen existing populations by increasing the amount of land and plants to support them."

As Andy and I talked, I became uncomfortable. What if he drove up the coast to my gardens in Cambria and found my patches of alien fennel festooned with swallowtail larvae, and hollyhocks bristling with feasting painted lady caterpillars? Would he think less of me if he knew that my groaning arbor of passion vine is nearly always the center of an orgy of feeding fritillaries?

I decided to confess it all and let him know that my gardens are a virtual crazy quilt of both local natives and introduced species— mute testimony to a sometimes catholic passion for greenery. He waited until I had unburdened myself and said, "My own garden is a mixture too. The front yard is mostly natives with some remnants of prior landscaping jobs. My backyard has veggies and fruit trees, exotics and natives. I'm just saying that people shouldn't go to their nursery and just buy whatever is there. They need to really make some conscious decisions about what species they choose to plant."

Calderwood's philosophy seems to dovetail with that of author Robert Michael Pyle, who wrote the invaluable *Audubon Society Handbook for Butterfly Watchers*. In his informative book, Pyle cited the importance of the native plant movement, and recommended that the nearer a piece of land is to its "natural state," the more one should emphasize landscaping with natives. Pyle acknowledged that most of us don't have the opportunity to live on an unspoiled piece of land, and so he advocates a judicious use of exotics in order to attract and keep butterflies in our home gardens.

After I spoke with Andy, I picked up my copy of L. Hugh Newman's classic book, *Create a Butterfly Garden*. Newman writes that many caterpillars feed on plants that are not "normally" grown in gardens, such as wildflowers and weeds. He stresses that nowadays our gardens are just too trim and tidy and that if you want to make your garden a "home" for the butterflies you just need to relax a bit and "allow a few weeds to flourish."

I like this laissez-faire gardening attitude. Newman's words of wisdom coupled with Julian Donahue's comment, "A lazy gardener is one of the best friends of wildlife," leads me to believe that I may have found my gardening niche. I'll just relax and read some of my collection of the *Butterfly Gardeners' Quarterly* newsletter that I've been saving for a rainy day...and I won't feel guilty. I know that outside, thanks to a little tolerance and benign neglect, the caterpillars will trade me a season of flashing, rainbowed wings for a few munched leaves.

HARVESTER

Faeries in the Fuchsias

"Fireflies are frequent guests;
Frogs stop by to eat and rest.
Starry flowers lure the moths
Who sip and savor sugar-broth."

—SHARON LOVEJOY,
Roots, Shoots, Buckets & Boots

When I was a child, I thought that faeries flourished in our fuchsias and tiny monsters lurked among the jasmine tobaccos and tomato vines. I shared my beliefs with my best friend, and we spent countless summer hours together in search of these diminutive garden dwellers.

On sunny mornings we crept stealthily into our neighbor's yard and picked patiently through tangles of leaves and vines. When one of us discovered a "monster" worm, we cooed a soft, mourning-dove-like sound to alert the other. Then, with much ceremony and mock bravery, we touched the worm with a twig (never with our fingers) and watched it rear its head menacingly. Our departures were always swift and unceremonious. We were sure that this monster was just an instant away from attacking us.

As the sun dropped behind the hills and the welcome coolness of evening settled into our valley, we stretched out on the prickly carpet of St. Augustine grass and awaited the arrival of the faeries. A soft purring noise signaled their approach long before we saw them. In the gathering darkness, we watched as they hovered briefly beneath the pendants of fuchsia blooms and then darted into the white trumpets of petunia that edged my mother's brick planter boxes. By the time the first wishing star appeared in the sky, these visitors were gone, and we were headed home for the night.

Many years later, my young son Noah informed me that we had faeries in our garden. We were sitting outdoors near a thick hedge of fuchsias. "Can you see them?" he asked as he pointed toward the hedge. Within a few inches of Noah, a large fawn-colored moth with white stripes and patches of pink on its hind wings hovered below the fuchsias. As we watched, it dipped in and out of the blooms with its long, tubular tongue, seeming more like a hummingbird than a relative of butterflies.

The next day we visited the library and found a color photograph of "our" faerie in an insect field guide and learned that it was a white-lined sphinx moth (Hyles lineata), a member of the family Sphingidae. Below this photograph, an illustration of my childhood "monster" was captioned, "Larvae of sphinx moth, also called hornworm or unicorn worm for the horn or spine on the posterior end.

Though formidable in appearance, the spine is harmless and is apparently only for show, to discourage hungry predators." My monster and faerie were one and the same, both beauty and the beast.

A few species of hornworm are considered pests and may be destructive to crops, but in the southwestern deserts the caterpillar of the White-Lined sphinx moth is an esteemed food of the Native American Tohono O'odham tribe. In his book *Cultures of Habitat*, scientist Gary Paul Nabhan describes a conversation with tribal member Remedio Cruz about preparing the fierce-looking critter for a snack. "We toast them in lard in a frying pan so that they taste like popcorn. Or sometimes we put them on a stick over a campfire, then string the cooked ones up like a necklace. Old ladies used to wear strings of them around their necks and snack on them like candy," Cruz said. Personally, I'll stick to pearls and real popcorn, and content myself with watching the caterpillars eat, instead of eating them.

Worldwide there are about 1,100 species of sphinx moth, with 125 in North America. The upright posture of its caterpillar inspired the eighteenth-century naturalist Carl Linnaeus to name it for its resemblance to the stately Egyptian sphinx. Because of a camouflage principle known as countershading (it is darker on the side normally turned toward the light), the caterpillar resembles a flat leaf and can be somewhat difficult to distinguish.

Often the tattered remains of a tomato vine and a pile of droppings are the telltale clues that you have a tomato hornworm (*Manduca quinquemaculata*) on your plant. Caterpillar versus plant, soul food versus real food: For me it is not a difficult choice. Although the taste of a ripe, sun-warmed tomato is one of my favorite gustatory pleasures, the sight of a graceful sphinx moth moving from flower to flower in the garden satisfies me even more.

The voracious hornworms grow rapidly and molt, or shed their skins, five times before dropping from their host plant and burrowing into the earth to pupate. Entomologist J. B. D. Meeuse describes the chitin-clad pupa of a sphinx moth as a "miniature Greek vase or jug"—some have curved, exterior "handles" that encase the long tongue. If you uncover one, examine it closely, then gently cover it with soil, place a rock or stake near the spot, and check daily for the emergence of the moth. You may be lucky enough to witness its debut as it slowly pumps fluid into its budding wings.

In *Seasons in the Desert*, naturalist Susan J. Tweit writes about these "unusual creatures," the sphinx moths. "With stout, furry bodies, long, narrow wings, and wingspans from two to six inches, these improbable beings look like a cross between bats and hummingbirds." Because of their characteristic flight pattern and resemblance to a host of other creatures, these moths boast a multitude of nicknames. The French refer to them as the sphinx-*moineau*, or sparrow moth, for their swift birdlike flight. They are also called bee moth, hawk moth, hummingbird moth, lady-bird moth, "my-lord-long-tongue" (for the long, coiled tongue that is sometimes twice the length of their bodies), and "lined evening-lover," though many of these moths zoom about during daylight hours.

The heavy-bodied sphinx moth, which can weigh as much as a small hummingbird, must maintain a body temperature of nearly 100°F to achieve its rapid, hovering flight. On chilly days, these moths are able to raise their internal temperature by warm-up exercises that noted zoologist Dr. Bernd Heinrich describes as "wing-whirring, a process analogous to shivering." Through a series of scientific experiments conducted in the early 1970s, Heinrich discovered that not only are the large, fuzzy moths able to generate

• Welcome the Sphinx •

Plant an array of fragrant, tubular plants to attract sphinx moths. Evening primrose, nicotiana, fuchsia, petunia, honeysuckle, polyanthum jasmine, garden phlox, and four o'clocks are proven winners in my gardens.

If you find a hornworm feasting on a tomato or other small larval host plant, cover the entire plant with cheesecloth to form a large tent. Weight the outside edges of the cheesecloth with a circle of heavy stones. Doing this will exclude the hornworm from other plants. When the hornworms disappear, you'll know that they have dropped to the ground below the plant and burrowed into the soil to pupate. Be sure to avoid tilling or disturbing the soil in this area.

Never use pesticides, herbicides, or fungicides in a pollinator-friendly garden.

heat, they are also able to dissipate a buildup of heat by using their abdomen as a sort of radiator.

Naturalist Tweit mentions that most of these fast-flying moths, with twenty-five to thirty wing beats a second, fuel their high-energy metabolism with flower nectar that is approximately 60 percent sugar, "twice that required by bees, half again that needed by butterflies."

At twilight many of the evening flowers on my small porch open, lift their heads to the sky—a process called nyctitropism—

and release their perfume. Sphinxes unerringly follow this invisible pathway of fragrance to the shining, pale blossoms. Poised gracefully above the awaiting blooms, the moths hover in midair, uncoil their long tongues, and sip the plant's sweet syrup.

What may seem like an "all take and no give" relationship on the part of the moths is just not so. Flowers produce nectar only to lure pollinators to their blooms. It is a sort of insurance policy and a floral reward system that ensures cross-pollination, a diverse gene pool, and abundant seeds and fruits.

As a sphinx moth sips at a blossom, a life-giving cargo of pollen dusts its proboscis, body, and wings. When the moth moves from plant to plant, the pollen from the male stamens of one flower is transferred to the female stigmas of another. The lightning-swift sphinx moths have been observed pollinating nearly 200 flowers in just under seven minutes. I can't think of a gardener in the world who wouldn't welcome that kind of assistance.

Although I am grateful for the pollinating power of the sphinx moths, I must admit that I would love them for their beauty and grace alone. Some gardens offer more than a harvest of bushels and bouquets. Just bless me with a raggedy tomato plant inhabited by monsters and a hedge of fuchsias alive with the sky dances of the faeries.

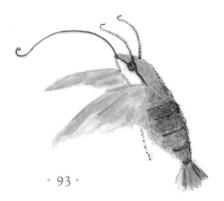

A Lap Full of Butterflies

They are a mystery to me, the burnished copper monarch butterflies that share my California and Maine gardens. For twelve months of the year, I work alongside them and watch them feed, mate, inadvertently move pollen from flower to flower, and glide like hawks across the skies. When the sun shines through their delicate, black-veined wings, they look like intricately leaded pieces of colorful stained glass. I often pause in the garden to watch and wonder how something so small and delicate can survive the rigors of a major migration. What ancient instinct unerringly guides them to the distant overwintering forests of their ancestors?

My grandmother called the monarchs "Mother Nature's wanderers," a well-deserved nickname. Every autumn the eastern population of monarchs migrates from their northern breeding grounds to warm overwintering sites. Some linger in the southern United States, but researchers estimate that from 100 to 300 million colonize the sacred oyamel fir (Abies religioso) forests in the highlands of Mexico. A smaller, western population of one to five million monarchs migrates from Canada and west of the Rocky Mountains to the California coast, where they congregate in thick stands of pine, cypress, and eucalyptus.

The monarchs' ability to find their way to their overwintering grounds astounds me. I am a fairly intelligent person (if you exclude my failures in algebra), and I get confused and lost just traveling the back roads of Maine. Yet the tiny, three- to four-inch monarchs, equipped only with a large ganglion, or nerve bundle, in their heads, are better navigators than I.

One tagged in New England was discovered in Michoacan, Mexico—a distance of 2,900 miles.

Although I have never seen the aggregations of millions of monarchs found in Mexico, my son Noah and I have spent some magical days observing a nearby California population. One sunny afternoon we walked to our favorite secret place in a grove of eucalyptus and spread an old picnic blanket in a clearing. We stretched out side by side, and looked up at a sky the color of my Gram McKinstry's 'Heavenly Blue' morning glories. Above us hundreds of monarch butterflies circled lazily or spiraled down the shafts of sunlight like falling autumn leaves.

We sat up back to back, and watched as a few of the monarchs cruised just inches from our faces. One male, identified by the small, dark spot of scent scales on his lower wings, followed closely behind a female. Hidden in the male's pouched scent spot is a chemical cocktail of pheromonal perfume secreted by abdominal glands. As the pair mates, the pheromone diffuses, wafting wordless messages to passing butterflies.

Long chains of monarchs, layered clapboard style one above the other, hung from the branches and leaves of the surrounding trees. Scientists believe that this mass clustering and layering protects the butterflies from predators, the ravages of rain storms, and desiccating winds.

Though the overwintering sites of the California coastline differ from the forests in Mexico, they share some similarities. Both have a dense canopy of sheltering trees, a mild climate (monarchs cannot survive prolonged freezing temperatures), and a plentiful supply of fresh water and nectar plants. Researchers have discovered that the removal of just one tree in a monarch roosting site can wreak havoc, altering wind patterns and exposing the sensitive butterflies to deadly temperature changes.

Late one afternoon a thick fog threaded its way through the grove, and Noah and I watched as the monarchs collectively shivered on their branches. A small, tight clump of butterflies fell to the damp ground and lay there unmoving. Monarchs need an air temperature of at least 55°F to fly; below that they are immobilized and become easy prey for birds, rodents, wasps, spiders, ants, and domestic cats. I knew that if I left the cluster on the ground they would perish. I lifted my denim skirt to form a cradle, made a nest of pungent eucalyptus leaves, and gently placed the monarchs inside. We walked to the sunny western edge of the grove and sat down on a sandstone boulder. I had a lap full of butterflies that we watched for nearly an hour.

▪ How to Attract Monarchs ▪

- Plant large swaths of native *Asclepias* (butterfly weed) in your home garden to attract egg-laying females and to feed developing caterpillars. You may observe as many as four or five generations of monarchs during a single season.

- Encourage a wild spot of native asters and goldenrod to flourish in your yard. You'll be amazed by the flurry of activity surrounding these maligned plants. Monarchs, skippers, and an array of other butterflies and bumblebees will thank you.

- Cluster nectar sources such as butterfly bush, heliotrope, zinnia, verbenas, including V. *bonariensis*, pentas, asters, Mexican bush sage (*Salvia leucantha*), pineapple sage (*Salvia elegans*), and goldenrod. This smorgasbord of flowers will sustain the monarchs throughout the changing seasons.

- Do not apply insecticides, fungicides, herbicides, or rodenticides on or near any butterfly nectar sources or caterpillar host plants. (Bacillus thuringiensis, Bt, is a killer.)

- Abandon your need for garden perfection and construct some small mounds of brush and twigs beneath trees. Butterflies will seek shelter inside the mounds during windy, cold, or rainy days.

At first the monarchs lay there, legs in a tangle, wings still, and stared at us with their immense compound eyes. Butterflies can see in all directions except right below their bodies, so we knew that they were aware of us. We offered a small branch of eucalyptus flowers to a male, who shifted slightly then moved languidly across my skirt.

Butterflies have the enviable ability to taste with all six of their feet, and their taste sensors are about 2,000 times more sensitive than ours. When the monarch's jointed black legs touched the blossom, his long coiled proboscis unwound automatically. He probed the bloom, angled and curved his tongue slightly, then leisurely sipped the nectar.

The sunlight absorbed by my denim skirt, coupled with the heat and light on their darkly veined wings, must have jump-started the butterflies back to activity. I could tell by their movements that they were reviving. The monarchs separated, opened and closed their wings, and began waving their antennae like enthusiastic orchestra conductors. Butterflies use their slender, club-tipped antennae to stabilize their flight (though their flight looks anything *but* stable) and to sift through a landscape of invisible stimuli. Their duo of deceptively simple-looking appendages glean information about possible predators, rivals, mating prospects, and food sources.

The butterflies climbed slowly from my lap and onto my legs and knees. One by one they left, ushered by a warm ocean-bred wind into the dark recesses of the eucalyptus grove. I stood up, smoothed out my rumpled skirt, and grabbed my son's freckled hand. Sharing my lap with the monarchs gave us a rare and intimate glimpse into their lives, but they are still a mystery to me.

Afternoon with a Spider

*Spiders should never
be taunted,
Maligned, abused, or
scorned—
They're hard at work in
our gardens—
Just look how that web is
adorned.
The filigreed tatting and
dewdrops
Tell tales of a
hardworking sprite—
An assortment of bugs,
Some mosquitoes, a grub,
Don't spiders get time off
at night?*

—SHARON LOVEJOY, *Hollyhock Days*

An unexpected visitor arrived on our porch stairs one afternoon late last summer. I nearly knocked her off the steps as I opened the screen door and headed outside to tend my container garden. Stretched between a string harp of 'Heavenly Blue' morning glories and a pot of verbena was an intricately woven web inhabited by an elegant black-and-yellow argiope spider (A. *aurantia*).

I stopped just inches from the handsome argiope, who sat head downward in the center of a web that was dotted with unrecognizable victims swaddled in silk. This opportunistic arachnid had slung her domain directly in the opening to my pollinator garden and was benefiting from the diverse insect population.

The spider took cover in the thick curtain of morning glories as I brushed an outermost strand of the web with a sprig of lavender. "I would never hurt you," I assured her and could almost hear my Quaker cousin Margaret Macdonald chant the old song, "If you want to live and thrive, let the spider run alive." She always ended her performance with the caution, "A good Scotsman never wastes anything, not even a spider's life." I'm not a Scot, but I didn't intend to do anything to harm this welcome guest.

I retreated to my worktable and grabbed Lorus J. Milne's *National Audubon Society Field Guide to North American Insects &*

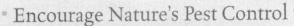

Encourage Nature's Pest Control

- Spread a thick layer of mulch over your garden beds. Spiders seek the shelter, warmth, humidity, and supply of garden critters beneath such a blanket.
- Do not spray pesticides on your gardens. Use the old "seek-and-squeeze" method or turn to an ecofriendly garden product.
- Set terra-cotta pots bottoms up throughout garden beds. Spiders enter the drainage holes and take up residence in the pots. Earwigs, sow bugs, and other unwanted guests hide in the pots and are consumed by the spiders.

Spiders, a pen, sketchbook, and magnifying glass, then slipped quietly out the door. The spider returned to her web and ignored me as I slid an old porch rocker beneath the morning glories, set my supplies on a bench, and settled in to share my afternoon with her.

The word spider derives from the Old English *spithra,* meaning "spinner," and the argiope is one of the best. The large web, backlit by the lowering sun, looked like an elongated, spoked wheel tatted with concentric rings of sticky spun silk. In the center of this nearly perfect doily was an erratic zigzag of webbing called a stabilimentum, which looks like a rickety ladder. This odd patchwork of shiny silk earned the argiope the nickname "the writing spider," because some people believed they could read letters and messages woven into its irregular framework. (It is always fun to share this story with children and to help them look for words or initials hidden somewhere in the spider's magical ladder of letters.)

Scientists once thought that the stabilimentum was a reinforcement to strengthen the web and protect it from damage. Others believed that the thick pattern of silk, which resembles the markings on the abdomen and legs of the spider, might serve as camouflage. Recent research by scientists in the United States, however, indicates that the shining, bright threads actually flash a warning "advertisement" to approaching birds, which often fly through unseen webs and destroy them.

A damaged web spells double disaster for an argiope; she depends on her sticky macramé for both food and lodging. A tattered structure no longer seines insect meals and requires an expenditure of time, energy, and protein to rebuild. Argiopes are the premier recyclers, ingesting their old web strands and reusing them for new construction. Web fragments carried away on the wings or bodies of birds cannot be recycled, and precious protein must be used to spin a new one.

The porch darkened as the sun dropped behind Miles Mountain. Inside the cottage, Jeff lit the Aladdin lamp on our dining room table. Within minutes the window was filled with moths, and the argiope's web vibrated with trapped victims. Although web-building spiders have poor eyesight, their sense of touch is highly developed. As each moth hit the sticky web, the argiope used her clawed feet to pluck the thin silk as deftly as a harpist. She felt, rather than saw, her trapped meal. The spider, who had straddled the web for hours without moving, now sprang into action. It took only seconds for her to grab each moth and shroud it in a tidy bundle of silk.

In E. B. White's beloved classic *Charlotte's Web*, Wilbur the pig bemoans the fact that his spider friend Charlotte is a bloodthirsty bug eater. Never one to mince words, Charlotte quickly sets him straight. "I live by my wits," the well-spoken arachnid informs Wilbur. And she adds, "If I didn't catch bugs and eat them, bugs would increase and multiply and get so numerous they'd destroy the earth."

Sheila Daar and Bill and Helga Olkowski, authors of *The Gardener's Guide to Common-Sense Pest Control*, concur with Charlotte. They note that Chinese research shows that arachnids are responsible for about 80 percent of the biological control in a garden. The authors state that "spiders are among the most important predators of insects, and their role in controlling insect pests is often underappreciated by humans."

Willis John Gertsch, author of *American Spiders*, credits arachnids with eradicating bedbugs in refugee camps, bugs in grain bins, caterpillar pests on coconut palms, cotton worms, gypsy moths on trees, and more. Gertsch calculated that a field near Washington, D.C., was home to approximately 64,000 spiders. It doesn't take a mathematical genius to deduce that if every spider in that field ate ten bugs each day (and that is a conservative guess), the number of pests consumed would be staggering.

Darkness fell and the Maine mosquitoes that survived the argiope's well-situated web soon found me. I picked up my book and drawings and stopped to bid my guest faretheewell. Thinking again of my cousin Margaret, I spoke the words that were printed on her meeting house wall. "Thee is welcome here, Friend," I said to the spider, and added, "Keep up the good work and I'll see you in the morning."

Night Life

"I like the warm, dark summer night,
When fireflies burn their golden light,
And flit so softly through the air,
Now up, now down, now over there!"
—ELIZABETH JENKINS

As the evenings soften and days lengthen, I am reluctant to end my time outdoors. I feel the way I did as a child when, despite my mother's calls to dinner, I lingered up in the branches of my favorite tree and wished on falling stars.

The thin sliver of time between sunset and night is my favorite. I always set aside those minutes to pause and observe the changing of the guard, from the familiar inhabitants of the day to the hidden, secretive members of the night shift.

Here in Maine, down the bay near Birch Island, an osprey is circling with the last fish catch of the day grasped firmly in her talons. The dragonflies are zig-zagging through their final insect patrol along the granite ledges, and disappear as the bats begin to comb the sky for their suppers.

As the gathering darkness erases the islands and the distant shores of Pemaquid Point, the 'Tina James' Magic' evening primrose (*Oenothera glazioviana*) captures my attention. I can't let the slow, golden dance of its opening flowers go by unnoticed. I find the most mature of the swollen, pink-tinged buds and watch for any perceptible movement. Soon, the bases of the blooms quiver slightly, and the pleated petals bulge out through the green sepals like the fabric of a partially opened umbrella. Within seconds, the sepals open completely, the pistil protrudes, and the four yellow petticoats of bloom unfurl.

I could stand here by the primroses until the sphinx moths begin their evening nectar stops, but some last-minute outdoor chores beckon, and I need to take a quick walk through the garden. Last spring, on such an evening walk, Jeff and I made a discovery that convinced us never to alter any area of our garden without first exploring it by both day and night.

The unruly, green shoulder of plants sprouting along the berm of our driveway made a horrible first impression of our Maine cottage. We decided to weed the area and landscape it with an array of native shrubs. On our final assessment walk the evening before we were to clear the land, a magical discovery changed our minds. Scattered throughout the untidy ribbon of greenery, like a necklace of shining jewels, were dozens of glowworms, the luminous, larval offspring of fireflies.

Fireflies are members of the large Lampyridae (from the Greek for "shine") family that comprises more than 2,000 species worldwide. They deposit their eggs in moist soil in wild patches of land and the forgotten, overgrown areas of gardens. The untamed stretch of ground beside our driveway obviously met all of their requirements.

The flattened, segmented glowworms are voracious eaters with an appetite for many of the critters I try to keep out of my vegetables and flowers. A late-night foray into the garden may catch the glowworms in the act of devouring larvae (sometimes even their own), mites, slugs, snails, cutworms, and, alas, sometimes even my beloved earthworms. But, I forgive the transgressions of this silent cadre of earthbound eating machines. I know that while I am sleeping peacefully they are eating some of my garden's worst problems.

The glowworms' season of peace and plenty ends at the approach of cold, when they burrow underground to overwinter. Come spring, when the weather warms and days lengthen, the glowing larvae will reappear. Some will feed throughout the summer. Others, at the end of their one- or two-year cycle, will feed for a short time and then disappear into snug, pellet-lined soil chambers to undergo the final transformation from glowworm to firefly.

This past spring the bedraggled berm along our driveway was studded with a population explosion of glowworms that made it look like a part of the star-sugared sky. Now, just a few weeks later, the same area shelters scores of female fireflies clinging to the grasses.

Above the grasses, through the bayberries, and past the pale trunks of the birches, the males circle, hover, and flash in search of

• Welcoming Glowworms, • Fireflies, and Luna Moths Into Your Garden

- Set aside a disorderly plot of ground for the wild things.
- Allow a mulch of leaves to remain on that plot throughout the seasons.
- Rake and cultivate sparingly or you may injure the overwintering pupae.
- Abandon the use of pesticides, herbicides, and fungicides.
- Use organic fertilizers.
- Plant native trees to provide larval food for luna moths.

the answering light of the hidden females. Locating a mate in the blackness of night is no easy feat.

The females, who usually remain in one spot, may totally ignore a male or glow faintly in answer to his flashing signal. But, if they are attracted to a male's flash, they will send out a quick, bright, answering flicker. The male responds and a cryptic dialogue of light between the two leads the male to the female.

Recent research by Ohio State University entomologist Marc Branham indicates that females are most attracted to and stimulated by the males with the fastest flashes. When a female spots such a series, she fires off a string of twinkling responses as bright as a neon sign at an all-night cafe. The lightning retort lures the male out of the sky and guides him to the receptive female.

As I walk through the birches, I can see a crescent moon rising over the point, its image mirrored in a slender, wavering thread of silver across the dark water of the bay. Up on the porch, the light from my drawing table is shining through a window and a celadon green luna moth (*Actias luna*) moves slowly up and down the glass, looking like a fairy who was issued an oversized set of gauze wings. The luna moth is astonishing and actually looks astonished, with its two pairs of owl-eye wing spots. This five-inch beauty has long, trailing "tails" that look like they were applied as an afterthought. I wonder how many times that filmy train has ended up in a surprised predator's mouth, saving the escaping luna's life.

My *National Audubon Society Field Guide to North American Insects & Spiders* states that the rare luna moth has only vestigial mouth parts and does not feed during its adult life. Though the moths don't feed, the tattered leaves on our oak and birch trees are mute testimony to the appetites of the fat luna caterpillars, who also feast on the foliage of walnut, hickory, willow, maple, beech, alder, persimmon, and sweet gum trees.

The fallen leaves under the caterpillar's host trees eventually become the protective camouflage for the cocoons of the pupating lunas. Amy Bartlett Wright, who wrote and illustrated the simple, informative *First Guide to Caterpillars*, mentions that if you are ever able to find one of the well-concealed cocoons, you may be able to hold it next to your ear and actually hear the pupa moving inside its thin winter quarters.

As I climb the porch steps I can see Jeff composing a makeshift picnic supper. I guess my short evening walk has leaped the boundaries of our normal dinner hour and crept into the night.

The chores I set out to do were never accomplished. In fact, I've forgotten exactly what they were. What I do remember is the brief cotillion of the evening primrose and the firmament of fireflies sparkling in the grass and inky, black skies. And an unlikely, giant green moth that looked like it was costumed for a midsummer celebration. All of this nightlife and more from a garden left untamed and a "landscape" never planted.

Birds, Bugs, and Benefits

"I noticed that phoebes will build in the same recess in a cliff year after year. It is a constant thing here, though they are often disturbed.... It takes us many years to find out Nature repeats herself annually."

—HENRY DAVID THOREAU'S JOURNAL, MAY 5, 1860

My thick book of North Carolina folklore says that "it is good luck to have birds come into the house." This old belief is found in many cultures, from the Penobscot Indians in Maine who believed "a little bird entering a house or tent is good luck" to Russian settlers in California who thought that "a bird flying through the house will bring luck."

Let me expand on this a bit and add that it is good luck to have birds in your garden and near your house. I am speaking from the perspective of someone who shared an intimate spring and summer with an audacious Eastern phoebe (*Sayornis phoebe*), a small, bewhiskered, buff-colored flycatcher.

The seven-inch bird flew into our lives in early May and claimed the buggiest, most sheltered area of our Maine cottage as her own. In the beginning she spent her daylight hours on a scraggly branch of spruce that overhung the busy pathway leading to our door. No amount of noise or commotion ruffled her. She ignored us and sallied forth from her perch, plucking insects from the ground and noisily snapping moths, mosquitoes, and mayflies out of the sky.

After a week of eating and systematic exploration of our porch beams and eaves, the phoebe settled on a homesite. For two weeks I watched as she carried mud, fibers, and grass to the top of an electrical box high under our roof overhang. She slowly constructed a four-by four-and-a-half-inch cup-shaped nest encrusted with an assortment of lichens and moss that she "plastered" to its sides.

When the nest appeared finished, I anticipated the arrival of the male phoebe, whom I often heard singing in the birches. "Fee-be, fee-be, fee-be," he called insistently in a rising and falling lilt hundreds of times a day. His repetitive "name-song" must have been a powerful temptation, for our phoebe disappeared and wasn't seen on her spruce branch or near her nest for days. I was worried that she had met some terrible fate.

On a Friday morning, as I headed off to do our weekly shopping, I glanced up toward the eaves. She was back and sat atop her nest like a lid on a teapot. I turned around and slowly closed the screen door. Inside the cottage I searched through bookshelves and located my *Peterson Eastern Field Guide to Bird's Nests*. Author Hal Harrison's field work indicated that a typical Eastern phoebe clutch numbers three to six white eggs; the last one or two may be dusted with freckles. He explains that incubation is done by the female alone and that she usually tends two broods in one season. I closed the book and went outside to check on the phoebe. Her muted, pussy-willow-gray feathers melded into the lichens, moss, and mud, making it difficult to tell where bird ended and nest began. I knew that under her soft, warm breast a small miracle was unfolding.

Winds and rain swept our rocky point of land and buffeted our cottage, but the phoebe remained snug and dry under the eaves. A fox appeared each evening and stared up at her; the phoebe stared back. A squirrel hung upside down from the edge of the roof and looked longingly at a meal just out of reach. The phoebe fluffed her feathers over the edge of the nest and closed her eyes. Although Jeff moved a ladder and paint bucket inches from the nest and painted carefully around it, the phoebe never moved.

Early one morning I sat on the steps of our porch, sipping my coffee and looking for the bird. It was the first time in weeks that she hadn't been right outside our door. After a few minutes she flew into sight, perched on her spruce branch, dipped and fanned her tail repeatedly (a typical phoebe family trait), and turned to face me. She looked comical. Moth wings and other unidentifiable insect parts bristled from her bill. She flew up to the nest and a faint chorus began to sound as she poked her head into the mossy cup and delivered a meal to her newly hatched brood.

The young grew and the nest site became messy and crowded. The phoebe was constantly "on the wing" making food forays. Although I could hear the male's call nearby, I seldom saw him near his mate and never observed him helping her with her dawn-to-dusk feedings. She kept up her frantic schedule for more than three weeks, literally vacuuming the sky and ground for insects. I was astonished by the volume of critters she delivered to her hungry, rowdy clan. For a few days I tried to keep track of the winged harvest, but the bird outdistanced me. By the time I noted a delivery in my journal, and before I could do anything else, the tireless worker returned to tuck yet another morsel into a gaping mouth.

I called Kimball Garrett, an ornithologist at the Los Angeles Museum of Natural History, to talk about the phoebe family's food preferences. I wanted to determine whether they were eating garden "pests" or the good guys. Garrett mentioned that up to 60 percent of a phoebe's diet is comprised of Hymenoptera—ants, bees, wasps, and sawflies—which are considered beneficial in a garden ecosystem. A smaller portion of the diet consists of grasshoppers, crickets, caterpillars, moths, mosquitoes, and beetles, some of which are considered "harmful" insects.

Garrett's information was consistent with an 1885 United States Bureau of Biological Survey study of the eating habits of wild birds. In the exhaustive project conducted by the Bureau, the stomach contents of sixty thousand birds of over 400 species were examined and analyzed. The resultant charts are fascinating and detailed.

Scarlet tanagers were observed in the field eating thirty-five gypsy moth caterpillars per minute, Nashville warblers ate three tent caterpillars per minute, and a whopping eighty-nine plant lice per minute were consumed by one tiny yellowthroat.

When the stomach contents of a rose-breasted grosbeak were examined scientists found the remains of fourteen potato bugs. A downy woodpecker had consumed eighteen codling-moth larvae, a

What to Do to Attract Bug-Eating Birds to Your Garden:

Situate birdbaths throughout your garden, but do not set them near bushes that provide cover for cats. Keep birdbaths clean and filled with fresh water.

If possible, provide a running fountain or trickle of water into a shallow basin. Both the sound and the movement will attract the birds. (Avian misters are available in nurseries, and from Wildbirds Unlimited stores, and catalogs.)

Plant a diverse array of trees, shrubs, annuals, and perennials, and concentrate on reintroducing native species.

Use a protective layer of mulch to thwart weeds, conserve moisture, and entice bugs. You'll find lots of birds poking through and under your mulch for critters.

Avoid using pesticides, herbicides, and fungicides. Poisons will destroy the beneficial soil dwellers and may sicken or kill birds.

red-winged blackbird twenty-eight cutworms, a robin 270 larvae of March-flies, and a flicker 5,000 ants.

The quantities of harmful insects consumed amazed me, but the diversity of species eaten by the birds shocked me. I went back to the charts three times to recheck the numbers. Downy woodpeckers were observed eating up to forty-three species, horned larks sixty, flickers eighty-nine, wood pewees 131, robins 223, cardinals eighty-one, bluebirds 166, phoebes 121, and nighthawks an incredible 600 species.

When the statistics of both field and laboratory work were completed, Survey scientists were able to prove to nonbelievers that a healthy population of birds can be the best friend a gardener or farmer can have. Even without the findings of the scientists, our twenty-gram flycatcher had already proven that to me.

After the young, rumpled-looking phoebes left their nest, I expected things would ease up a bit for the mom, but they didn't. The four youngsters claimed branches up and down the spruce tree, quivered, shook their feathers, opened their mouths, and begged for more handouts—and the female dutifully fed them for a few more days.

One morning they were gone, and I missed them and all of the activity. The nest sat empty and in disrepair, and our freshly painted wall looked like a Jackson Pollock canvas. I heard a male calling in the distance and our female, probably worn to a nub, was nowhere in sight. Perhaps it was my imagination, but I thought that there were more mosquitoes than normal, and that a siege of insects was already attacking my struggling box of basil.

Ten days later the phoebe returned. Crusty gray lichens hung from her black, bristle-fringed bill. She flew up under the eaves to her disheveled nest and began the first of many repairs. By the end of the week, the nest was good as new and our wall looked even worse. Intuition told me that we'd better not repaint for awhile—it seemed like our good luck had returned.

The Good, The Bat, & The Ugly

Gentle weavers of the night sky are our most misunderstood garden helpers.

I was spending a few days with my friend Julie in the foothills of the Sierra Nevada. Julie's guest house was a small garden shed set just a few feet from her kitchen door. Before I settled into the soft bed, I extinguished the lights and pushed open the screenless window. The thick fragrance of night-blooming jasmine and roses, and a chorus of crickets, filled the room and comforted me to sleep. I dreamed of gardens brimming with old-fashioned flowers and awoke suddenly as something brushed lightly past my cheek. I sat up and bumped into something. My heart raced as I tried to orient myself to the strange surroundings. I was alone in a cottage in a blanket of darkness with an unknown intruder.

Slowly, deliberately, I reached above my head and tugged at the lamp chain. As the welcome light flooded the room, a small brown bat executed an acute turn and swooped away from my extended arm. I flattened myself against the mattress and thought about all of the horrible bat stories I had heard through the years. Though I love bats and have observed them from afar, I never had one as a roommate until this particular night.

For a few minutes I was privileged to watch one of the world's foremost insect catchers in action, just inches above my head. Moths lured by the brightly lit lamp were flying erratically through the window toward the light only to be picked off, one by one, by my roommate. I smiled as I nestled into the covers and listened to the bat crunching his way through his midnight meal.

Eventually I realized that the bat performance had to end, so I turned off the light, walked slowly over to turn on the porch light, and opened the door. Swoosh, and the bat was out of the room and busily scouring the night sky. I said goodnight, closed the door and the window, and crawled back into bed.

The next morning I wondered why, if bats have such great sonar, my evening visitor had run into me. I called the Los Angeles County Museum of Natural History and was told in no uncertain terms that bats have such powerful echolocation they can easily dodge a thread dangling in a darkened cave. Evidently, I had popped up in front of the bat so swiftly that both it and I were surprised.

This starlit evening shared with a small, furry, flying mammal led me on a trail of discovery. Always one for trivial facts, I took copious research notes on bats, possibly not important or necessary to the normal person's bank of knowledge, but invaluable in working with children who often find them frightening.

• Here Are a Few Things •
You Can Do, as I Have, to Put Out
a Bat Welcome Mat

- **Provide shelter from sun, rain, and predators.**
Shelter can be as simple as a large brush pile in an
out-of-the-way area, a stone dry wall, or a loosely constructed pile of 10- to 12-inch rocks that will not be
disturbed. Plant your dry walls or rock pile with an
evergreen vine to give further protection. If you have
dead trees on your property and they do not pose a
threat, leave them standing for the benefit of bats and
cavity-nesting birds. I have noticed that many bats
will seek shelter in exterior shutters. Recently, while
on a trip to Michigan, I found a wonderful pair of six-
foot-tall antique shutters and simply screwed them
onto the side of a sunny, southeast-facing wall about
12 feet above the ground. I made sure there was a
clear approach area so my eagerly anticipated bats
could easily swoop between the slats of the shutters.
- **Provide fresh, clean water.** The bigger the water
source, the better. Unlike birds, who can sit at a bird-

Bats, one of the most maligned and misunderstood of all our
garden helpers, are an absolutely essential thread in our fragile web
of life. The United States boasts forty species of bats, although sadly
with over 40 percent of their population at dangerously low levels.

bath and drink, bats need to drink on the wing, lapping up water as they skim the surface. Ponds can be simply constructed with butyl liners, or you can use a pre-formed plastic pond. Check your local nursery or your water garden catalog for complete instructions.

- **Plant a diverse habitat.** Select native plants, flowering vines, evergreen vines, herbs, night-flowering plants (moonvine, night-scented stock, nicotiana, jasmine, heliotrope, four o'clock, evening primrose, silene, dame's rocket, honeysuckle, and bouncing bet are just a few), grasses, trees, and shrubs. Think of your garden as sort of a layer cake of bat food. Start at the bottom layer with grasses, add a layer of flowers, another of sub-shrubs, some taller shrubs, and finally the top layer of trees. A finishing touch includes embellishing bare walls and fences with blooming vines and evergreen vines to provide shelter. This layered habitat will entice a parade of insects and moths that are, after all, bat snacks.

- Avoid the use of harmful herbicides, insecticides, fungicides, and pesticides.

Kids ask me, "So what?," and I carefully explain to them that bats are our best bug eaters, pollinators, and seed dispersers—and the midwives to our rain forests.

My Grandma Nonie called bats "flitter-mice" and for many

Bat, Bat,
Come under my
And I'll give you
a slice of
And when I bake
I'll make you a
If I am not mistaken.

---

years people believed that bats were flying mice or birds, but bats are flying fur-bearing mammals who produce usually only one pup a year. Bats belong to the order of mammals called Chiroptera, or "hand wing," and if you ever looked closely at a bat, you would see that its wing is really an elongated hand, with extended fingers connected by a thin, flexible skin. I love pointing out to children that their fingers would be an astounding four feet long if they were proportional to a bat's "finger."

If you become a seasoned bat watcher, you may have the rare good fortune of watching a mother bat doing triple duty: flying through the air, mouth open for echolocation, and nursing a clinging pup while catching dinner on the wing. One insect-eating bat can consume up to 500 insects per hour—a feat that helps to explain why the Chinese consider bats to be *fu*, a blessing to the garden.

"Crepuscular" is a great word that rolls around in my mouth like a handful of jawbreakers. Bats are crepuscular, emerging on the thin edge of twilight to feed and continue their beneficial bug work under cover of darkness. I love bats and want them to feel welcome and at home in my gardens.

In order for your garden to be "friendly" to a multitude of flying and ambling visitors, you'll need to relax your neatness standards a bit, and leave mulch and leaf litter on the ground. If you *must* be tidy, rake the litter into your beds. More habitat!

Bug-zapper lights do not kill mosquitoes and other insects that are not attracted to lights. Replace your zapper with a white incandescent bulb, then sit back and watch as bats zoom in and pluck darting insects from the sky as easily as you pick ripe cherry tomatoes in August.

Although bats are incredibly gentle and beneficial creatures, they are still wild. It is not natural for a bat to be on the ground. If you find a bat on the ground, do not attempt to catch it in your hands. Bats are fragile and easily injured, and there is a good chance that a grounded bat may be injured or ill. Quietly and gently cover the bat with a box and call the American Bat Conservation Society or Bat Conservation International for guidance.

Creating a garden for bats has added another dimension to my gardening. During daylight hours I never seem able to just sit, relax, and enjoy. The second I sit down on a garden bench my eyes see a weed that needs pulling or a plant to be deadheaded. But when the sun goes down in my bat garden, I sit and luxuriate in an india-ink darkness starred with night blooms and the meteoric forays of my *fu*, my tiny garden blessings.

Hummingbirds in the Garden

How to create a haven for these tiny,
colorful creatures

The short days of winter have settled quietly over my oak-sheltered gardens. A covey of russet brown quail carpet the ground under a feeder crowded with jays, towhees, white-crowned sparrows, finches, and pine siskins. Juncos, chickadees, and a family of clown-faced acorn woodpeckers vie for space at another feeder. Below my studio window I can watch the nuthatches and brown creepers bathing in terra-cotta saucers, and an Anna's hummingbird darting into a fountain's stream of clear water, while a magnificent red-shouldered hawk glides silently in for his daily bath and refreshment.

The garden hushes as the hawk bathes awkwardly in the huge old fountain. Only one critter seems unruffled by his presence—the tiny, fearless, and sometimes quarrelsome Anna's hummer, who calls this garden home twelve months a year. The hummer, a four-inch bundle of energy, repeatedly dashes into the hawk's face, scolding and stitching the air around him. Finally, tired of being harassed, the hawk climbs to the side of the fountain, shakes out his thickly-feathered wings, and resignedly coasts away as the hummingbird resumes his drinking and showering.

I am constantly enchanted and amazed by the antics of hummingbirds. I have spent years observing and drawing them for my personal journals. I've rescued them from untimely deaths, brought them out of torpor on top of a family heating pad, hand-fed them from a dropper, and replaced babies who have fallen from their crowded, lichen-laced nests.

All of my encounters with hummers have enriched my life. From a dazed female Allen's hummingbird, who lay quietly on my drafting table, I learned that hummers have long, flattened, tubular tongues, which are split into two smaller straw-like tubes perfect for sipping. Examining the long, delicate beak and the even longer tongue made it clear why hummers so love feasting at tubular flowers such as the salvias and trumpet vines. A flower long enough to exclude bees and other insects is simply saving its bounty for the visits of the hummers.

Imagine a metabolism so fast (its wings beat up to eighty times per second) that the hummer requires eating every ten to twelve minutes during the day. That is the plight of the hummer, who races from

• It's Never Too Soon •

Just follow these simple guidelines to begin planning the various ways to invite hummingbirds into your garden:

- Hummers love orange and red flowers, trumpet-shaped and single blooms.
- Salvias are a hummingbird favorite and a sure choice for a small space.
- Use a mix of annuals, perennials, shrubs, vines, and flowering trees to develop varied food sources.
- For close hummingbird observations, plant window boxes and hanging pots with fuchsias, nasturtiums, salvias, snapdragons, petunias, bee balm, native columbine, and coral bells.
- Plant a fence line with old-fashioned single hollyhocks.

bloom to bloom from morning to night to eat more than 60 percent of its body weight each day in nectar and insects. Early evenings, hummingbirds can be found probing the bark of a tree for small insects and grubs, plucking gnats from the air, and tanking up on nectar from flowers and sugar-filled feeders. The rightfully tired hummers, their crops filled with nectar to fortify them through the night, sleep 'til the morning when their frenetic, calorie-consuming lifestyle begins anew.

Gardeners from Europe and Japan who have never seen a hummingbird have written me envious letters asking for information and photos. They are astounded by some of the feats a hummer can

- Provide a dripping or misting water source. Hummers will return again and again to moving water.
- Avoid spraying plants with herbicides and pesticides.
- Buy a simple, easy-to-clean hummingbird feeder. Never allow mold to form in the feeder; this can kill hummingbirds. Wash and rinse thoroughly with hot water.
- Always use cane sugar in a feeder and follow the feeder manufacturer's recommended ratio of sugar to water. Never use sugar substitutes or honey: Sugar substitutes have no nutritional value for hummingbirds, while honey can ferment, causing their tongues to swell—a fatal condition.
- Consult your local libraries and bookstores for more information about hummingbirds.

perform—hover, fly backward, do aerial somersaults, live atop 15,000-foot mountains, migrate thousands of miles, and, in the case of the Ruby-throated hummer, travel 525 miles nonstop from Florida across the Gulf of Mexico to the Yucatan Peninsula.

Though there are 320 recognized species of hummingbirds, North American gardens can lay claim to only eight resident breeding species and eight visitors. The eastern to midwestern United States hosts that intrepid traveler, the Ruby-throated hummer, who arrives in the spring, mates, raises a family, and leaves as autumn approaches. The western United States is home to the Black-chinned,

Costa's, Anna's, Calliope, Broad-tailed, Rufous, and Allen's hummingbirds.

Bloom-filled desert gardens and riparian canyons in southeastern Arizona are often visited by Broad-billed, White-eared, Berylline, Violet-crowned, Blue-throated, Lucifer, Plain-capped Starthroat, and Magnificent hummers (the $5\frac{1}{4}$-inch Magnificent has wings that sound like a mini-helicopter).

My coastal California gardens are blessed with conversational Anna's hummingbirds year round and beautiful orange, red, and green Allen's throughout the spring, summer, and fall. Although their spirited presence in the gardens is common, it is never taken for granted. I have worked long, hard hours planning and planting for their visits. The hummers have now rightfully claimed this half acre of land as their own. They are here by invitation—a vibrant floral invitation that cascades over rock walls and flanks the pathways; a watery invitation that trickles from fountains to fill stone basins; a thorny and twiggy invitation that entices them to build their fragile lichen and cobweb nests on protective branches; a myriad of invitations that provide a welcoming habitat, abundant food, and safe shelter.

Your winter dreams and garden schemes can blossom into your own personal haven for hummingbirds. Though you may not take the hummers' presence for granted, you will be surprised to find that they may take *your* presence for granted. Accustomed to you working in your garden, they will play and bathe in a spray of water from the hose, sit on the flower sticking from your hat band, investigate a floral shirt, or perch precariously on a newly-filled feeder as you take it back to the patio. When you are world-weary and tired, turn to your garden—the hummingbirds will be there to greet you.

My Morning Alarmers

*"The chickadee is a noisy, restless little acrobat
as well as an educated musician, and his
appearance with a dozen of his fellows in the pine
tree near my cottage is the signal for a circus
performance with an orchestral accompaniment."*

—F. SCHUYLER MATHEWS, 1904

This blue-skied winter morning is my idea of perfection. The welcome sunlight shining through the kitchen window illuminates a row of copper pots and a colorful antique Staffordshire bowl inscribed with one of my favorite quotations, "I have lawns, I have bowers, I have fruit, I have flowers, the Lark is my morning alarmer." Outside in the gardens, my morning alarmers, a flock of chestnut-backed chickadees

(*Parus rufescens*) chatters sociably and flutters down to the empty feeders like drifts of falling leaves. The Cherokee people call the chickadees *tsikilili*, the bringer of news, and I think the news they bring this morning is that I am neglecting my breakfast duties. I grab the old honey tin filled with black-oil sunflower seed, slip on my clogs, and head outside to tend them.

Although winter along this stretch of the California coast is mild, all birds have a hard time surviving during this season. Insects are scarce, larvae are hidden under leaves or deep in the crevices of bark, and only a few shriveled berries and fruits remain. Fresh water, both for bathing and drinking, is as critical in winter as in summer, and shelter from the elements can sometimes mean the difference between life and death.

To help the birds through winter, Jeff and I plant dozens of native berry-producing bushes, vines, and trees, and fill large containers with annuals known for their copious seeds. We leave the plants untrimmed until the birds pick off all the berries and flowers.

Birdfeeders that meet every need abound in the gardens. We have platforms for quail, towhees, native band-tailed pigeons, woodpeckers, and scrub and Steller's jays. Fat pinecones, packed with my energy-boosting peanut butter blend (see box on page 132), hang from wires. Black-oil sunflower seeds fill cylinders tucked under the eaves, while long mesh bags of thistle sway beneath tree branches. The chickadees, my tamest and most vociferous companions, sample nearly everything and are always the first to arrive at the feeders as I stock them.

To provide natural shelter for the birds, we planted small copses of thick evergreens—some low to the ground for towhees and quail, others towering and widespread for mixed groups of song-

To Help Birds through Winter

- Provide fresh water daily and remove ice as necessary.
- Fill at least two kinds of feeders with seed (mixed and black-oil sunflower seeds), and hang them in a protected area out of reach of cats and greedy squirrels.
- Build a brush pile of twigs, branches, and rose or bramble canes. Birds will seek shelter inside the pile during the night.
- Hang nesting boxes under eaves and fill the floor of the boxes with cedar shavings 2 inches deep. Birds may use these boxes during bad weather and cold nights and then return to them for nesting.
- Make a supply of my high-energy peanut butter mix for hungry winter birds. This nutritious meal is as beneficial as a suet mixture and much easier to prepare. Chickadees, and thirty other species, will fly in to snack.

birds and an occasional hawk or owl. Protected, eastern-facing fences and walls festooned in thick tangles of vines attract flocks of birds seeking cover. As the first rays of sunlight pierce the lacy curtain of vines, the birds emerge like actors on a stage, fluff and preen themselves, and then head straight for the birdbaths and feeders.

Sometimes natural shelters do not offer enough protection during the worst of storms. A few winters ago, after an unusually cold night, I awoke early, put on a heavy bathrobe, and hurried outside to check for frost damage. The sun was just rising above the Santa

Lucia Mountains, and the yard was uncommonly hushed. As I walked along a pathway, I heard a faint rustling, then a muffled chattering as three chickadees emerged from a small wooden nesting box. They shook their rumpled feathers and bent tails and took flight. This early morning lesson in lodging taught me that birdhouses are a valuable refuge in the garden long before the spring nesting season.

I feed the birds all year because I love them and enjoy their companionable, lively presence in my life, but I also reap immeasurable benefits from them. Last year hundreds of the bright green caterpillars of the introduced white cabbage butterfly invaded a portion of my trial gardens. I resigned myself to the loss of my first planting of nasturtiums, since I will not use any pesticides and handpicking was an impossible task. Late one afternoon, I noticed two chickadees flying in and out of a nearby birdhouse and heard the faint, squeaky calls of their young. I peered through a spotting scope and watched as one of the parents—with a mouthful of squirming green caterpillars—landed on a twig directly in front of the birdhouse. He looked around to make sure that no predators were near, then flew straight through the entry and disappeared inside. Within seconds, he reappeared at the doorway and left just as the other parent arrived with another fat caterpillar.

The tireless chickadees kept up their assembly line of feeding until darkness settled. Over the course of three hours, they rid my gardens of several hundred of the destructive cabbage caterpillars. I wondered how the five-inch birds could consume such an enormous quantity of food. "Good day's work," I said aloud, as one of the chickadees flew past my window and slipped inside the birdhouse with the final meal of the day.

Late that night I curled up with a stack of books to do some research on the incredible appetites and attributes of my small garden helpers. One of the most informative books, a 1916 edition of *Bird Friends*, by Gilbert H. Trafton, contains the statistics of a massive bird study conducted by the Bureau of Biological Survey. Detailed field notes and the actual inspection of stomach contents revealed that the sheer number of insects that birds consume makes them an effective means of pest control. After witnessing what the chickadees accomplished in one afternoon, I consider that an understatement. Trafton wrote that birds, with their high body temperatures, rapid heart rates, and speedy digestive systems, process and assimilate food in less than two hours. Some birds eat enough insects to fill their stomachs a dozen times a day.

I tried to calculate the number of caterpillars that two adult chickadees and their offspring (typical nests contain five to eight young) might devour, and the numbers astounded me. Since nestlings require feeding every few minutes from dawn until dark, their parents could rid my garden of thousands of pests before the young fledge. I checked Trafton's index for any other mention of the birds and found a reference to the work of Professor E. D. Sanderson of Michigan. He wrote that chickadees in Michigan alone destroy eight billion insects annually. Enough incomprehensible statistics for one night, I thought, as I moved the pile of books to a bedside table and switched off the lamp.

For the next few months, I watched the chickadees raise two additional families in my birdhouse. I spent so much time with the adults that they became nearly tame. As soon as I walked outdoors with my tin of black-oil sunflower seeds, they arrived on the scene. For pleasure (and with a measure of eccentricity), I scattered

• Sharon's Super •
Energy-Booster for Birds

1 cup crunchy peanut butter

1 cup canola oil

4 cups yellow cornmeal (never delete this
    ingredient; it prevents choking)

1 cup white flour

(I sometimes add raisins and sunflower seeds)

Mix the ingredients thoroughly, then store them in the refrigerator in a covered container. Use a knife to spread the mixture into the scales of a pinecone. Hang the cone in a sheltered area out of reach of cats and squirrels.

sunflower seeds around the brim of my gardening hat. As I walked through the gardens, the scrub jays eyed me speculatively, but kept their distance. By contrast, the brave little chickadees dropped from the tree limbs onto the hat, snatched a seed, and retreated.

Sometimes, when I look at my feed store receipts, or fret over the money spent on birdhouses, I remind myself of the money saved by caring for the birds. I will never need to purchase pesticides or invest in a clock radio so long as the chickadees, my "morning alarmers," are content to be in my gardens.

A Thrush in
My Picnic Basket

*An array of berry-producing shrubs
and trees is a magnet for these birds.*

Their arrival in my garden is always quiet, as hushed as the dun-colored leaves that sift through the gnarled canopy of oak branches and drift to the ground. I watch for them, hoping to see their familiar sprint across the pathways and into the thicket of berry-laden shrubs.

The shy hermit thrushes (*Catharus guttatus*, for "spotted") that inhabit the forested grounds surrounding my cottage are easy to overlook. With their somber coloring and buffy, speckled chests, they vanish in the dappled shade, reappearing only when the flick of their wings and up-and-down pumping of their tails (their characteristic trademark) draws attention to them.

My tiny window-lined studio sits high above the gardens and the mossy fountain that is irresistible to the thrushes. They are constant visitors to this shallow four-foot-wide pool, jumping in and then flailing and flapping as though they are drowning. Sometimes they line up four or five in a row, like cars awaiting their washes, then plunge into the water one after another.

For years I watched these reclusive birds only from my studio, with the help of a spotting scope or binoculars. Other than photos in a bird book, I never saw them up close until the day I found an injured hermit thrush on the ground beneath a native currant. One of the bird's eyes was missing, and it appeared to be in shock. I knew that if I didn't do something quickly, it would die.

Nobody from our local wildlife rehabilitation center was available to help, so I grabbed Care of the Wild Feathered and Furred and skimmed through the text on bird care. To minimize shock, it said, maintain the bird's body temperature, and protect it from neighborhood cats. I needed a dark, well-ventilated container for the thrush. I scanned my potting shed for something suitable and found a lidded wooden picnic basket to use as its temporary home. I moved the basket onto my worktable and wedged in a small branch of native toyon, or California holly, for a perch.

The book offered a complete chart for feeding thrushes, and it took only a few minutes to prepare a small bowl of water and a crushed mush of berries, suet, ground beef, and raisins. I gently lifted the thrush into the basket, wished her (I use this term loosely as the male and female look alike) well, and closed the lid. I awoke before the sun peeped over the Santa Lucia Mountains and ran outside to

• Thrush Mush •

My thrushes enjoy raisin and bran cereals with some dried cherries, alpine strawberries, elderberries, blackberries, wild currants, holly berries, a dab of wet dog food, and chopped suet.

1. Lightly crush cereal with a rolling pin or heavy skillet. Mix in remaining ingredients.
2. Spread mix on a clean tray, or fill a bowl or hollowed half melon, grapefruit, or orange.
3. Situate feeder on a post in a clearing so cats can't sneak up on thrushes while they feed.
4. Keep the feeder clean. Empty unused contents every few days (I toss mine into the worm bin) and replenish with a fresh supply.

check on the thrush. Worried that she hadn't survived the night, I lifted the lid and found her perched on the toyon branch. She slowly tilted her head and gazed at me with her one eye, then sidled up a twig and onto the rim where she sat and surveyed my every move. I opened the potting shed doors so the thrush could leave whenever she chose.

During spring, I watched the little one-eyed bird as she foraged in my gardens. She taught me that thrushes are

• Fruit-Tree Feeder •

1. Fill a large terra-cotta pot with a quick-setting concrete mixture. For a natural look, stir in some pebbles and sand.

2. Before the concrete hardens, poke a dead multi-branched tree (size appropriate to container) into the slush.

3. Make hanging hooks out of short pieces of wire and attach them to the branches.

4. Hang clusters of grapes, branches of elderberries, holly, firethorn, cotoneaster, and toyon, and short fruiting blackberry canes from the hooks.

experts at eating-on-the-wing, plucking berries as they fly or hovering like kestrels (a.k.a. sparrow hawks) above their fare. They are also proficient stalkers, running along the ground in search of prey then stopping abruptly to strike at caterpillars, ants, spiders, beetles, grasshoppers, sow bugs, worms, and the occasional slug and snail (hurrah!).

The unobtrusive thrushes, overlooked most of the year, are easy to spot in the spring, when they perform their frenzied, circular flights of courtship. After mating, the female builds a small (three- to five-inch) nest, on the ground in the Northeast and in the low branches of conifers and shrubs in the West. Inside the compact cup of mosses, bark, and stems, she lays her pale blue oval eggs on a soft bed of fine roots, grasses, and animal hair.

• Plant These for the Thrushes •

Serviceberry (*Amelanchier*), bayberry, blackberry, blueberry, cherry, cotoneaster, native currant (*Ribes*), dogwood, elderberry, grape, holly, firethorn (*Pyracantha*), sumac (*Rhus*), and toyon, or California holly (*Heteromeles*).

During the thirteen-day incubation period, the male thrush feeds and guards his mate as she sits on her clutch of eggs. During his sentry duty, he sings a flutelike song that F. Schuyler Mathews, in his 1904 *Field Book of Wild Birds and Their Music*, described as that of the American nightingale. Mathews contended that the clear melody, which can be heard up to a quarter of a mile away, is unequaled in the world of bird music, and I must agree with him. Although I have never heard the hermit thrush utter more than short "chucks" and "chups" in California, in Maine I can count on him to sing down the sun.

Through autumn and winter, brilliant clusters of red berries provide exclamation points of color in my wild gardens, but by late

spring, the thrushes have stripped the bushes bare. I never begrudge the birds their harvest, for they always give so much more than they take. They are little gardeners par excellence, planting—albeit in some strange places—a fresh crop of berry bushes each year. Wherever the hermit thrushes go, they leave a green trail of life behind them...as fine a legacy as anyone could wish.

Getting the Blues

Easy ways to shelter bluebirds and purple martins

It was love at first sight when I saw the glossy purple martins plying the insect-laden air for their breakfast. I felt the same when a flock of twenty-two eastern bluebirds settled, like pieces chipped from the cerulean sky, into a grassy meadow below me.

I knew, as I sat for hours watching the birds, that someday I would find a way to shelter them on my own land. Now, my heart and mind have turned to a new garden in Maine, and our plans are being sown for habitats and shelters for everything from foxes to spiders.

The other night, as Jeff and I spread drawings and plans across our long kitchen table, we totaled the figures for all of the houses, roosts, and plants we hope to buy. I hadn't even mentioned my dream of a granite birdbath, and Jeff had already started worrying about our budget and other necessities. While he was fretting about food for us, I was focusing on food for the soul. Just as Imelda Marcos feels that you can't have too many pairs of shoes, I believe that you can't have too many plants or nesting birds.

At a nursery, we found some huge wooden and metal "apartment buildings" for the sociable martins and about six different styles of bluebird houses, but the prices were staggering. Even if we could somehow prune a bit from our budget, there was no way we could ever afford to house all of the critters I hoped to welcome. A search for low-cost, alternative housing that was suitable for these kinds of birds and pleasing aesthetically began in earnest.

At Longwood Gardens in Kennett Square, Pa., I learned how to solve my problem of housing both martins and bluebirds. While our friend Dave Thompson of Longwood's Education Department was escorting us through the "Idea" garden, I photographed and coveted the plump gourd birdhouses swinging gently from a tall gourd rack. He told me I should talk with the section's gardener, Bill Haldeman, who has been instrumental in developing and maintaining wildlife habitats throughout the grounds at Longwood. Thompson also said that if we were looking for successful, economical bluebird houses, we had to call Warren Lauder, a man he affectionately called the "Bluebird Man."

When we returned home from Longwood, I called Nichols

Garden Nursery in Albany, Oregon, for a few packets of seeds from their gourd collection, which includes the birdhouse gourds (*Lagenaria siceraria*), named for their plump, bottle shape. My hope was to grow enough of the gourds to provide housing, similar to the rack at Longwood, for a healthy population of martins.

I compared notes with Bill Haldeman about the cultivation of the gourds. He grows his gourds in full sun in the corner of his compost pile, where the constant moisture and rich, well-drained soil make the vines thrive. Usually, seven or eight seeds are planted one inch deep in small hills about six or seven inches high. The hills are spaced about six feet apart to give the rambunctious vines room to gallop and roam.

Bill cautioned, "If you use too much nitrogen, you'll get lots of growth, but no fruit. You want to pinch off some of the developing fruit and try to grow four or five big gourds. If you get some healthy ones with good, thick skin, save the seeds for next year's crop. Don't save thin-skinned gourds: They'll crack and be useless."

I planted my gourds against a sunny fence and barn wall in the Heart's Ease gardens. My first crop was seeded directly into a rich, deeply dug bed, but cool weather and an overabundance of slugs and snails destroyed it.

I soaked my second seed packet in a cup of water for two days and planted the seeds in biodegradable pots I made from newspaper. I kept the paper pots up on a bench in the full sun, and every evening I covered the seedlings with old jelly jars to keep them from being munched.

After the gourds were up and healthy, I transplanted them, pot and all, into the children's garden where they clambered and

• Grow Your Own Martin Houses •

What You'll Need:

- A packet of birdhouse gourd seeds (*Lagenaria siceraria*)
- Paper cups
- A sturdy trellis, tepee, fence or arbor
- Soak seeds overnight.
- Plant seeds in small paper cups (don't forget drainage holes) filled with sterile potting mixture.
- When seedlings show their first set of true leaves, they may be planted.
- Gently cut off the bottom of the cup without injuring roots.
- Plant the cupped gourd in a sunny (6- to 8-hours a day) area with rich, quick draining soil that is mounded into small hills about 8- to 10-inches high, spaced 6-feet apart.
- Water thoroughly (at base of gourd) every day.
- Pinch off some of the developing fruit to encourage big, thick skinned gourds.
- Allow gourds to remain on the vine until after a couple of frosts, or if you don't have frosts, until the vine browns and withers.
- Cut gourds from the vine (leave a few inches of vine attached) and dry them in a cool, sheltered area out of weather for about six months, until they rattle when shaken

Crafting the Martin house:

- Use a sharp knife to cut a 2-inch entry hole 3- to 5-inches up the side of the gourd.
- Clean out seeds and save some for next year's planting.
- Use a pocket knife to enlarge the bottom of the hole $1/4$-inch so it is pear-shaped.
- Use a $1/4$-inch drill bit to bore a series of holes around the top and bottom to provide ventilation and drainage.
- Drill a small hole through the top of the gourd (above entry hole) and string it with a plastic-coated wire hanger for mounting.
- Martins choose white-painted gourds over natural ones. Paint gourds with white exterior enamel, which will highlight the entrance hole and reflect the heat.
- Attach at least a dozen gourds at one foot intervals to a clothesline mounted on tall metal poles 20 feet above the ground in a large clearing (they need open airspace for their flights) at least 40 feet from buildings and trees.
- Place a protective metal shield or collar (to thwart squirrels, raccoons, etc.) 4 feet above the ground on the metal mounting poles.
- To discourage interlopers, don't hang your gourds until a month after Martins arrive.
- Don't outfit gourds with perches; the Martins don't need them, but other species will use them to gain access.

looped up the playhouse, over the walls, and into areas not meant for them. Pruning them encouraged branching and fruit production; the only problem was that I couldn't keep up with their exuberant growth. Next time I plant gourds I'll use trellising and old ladders for support and I'll only use half a packet of seeds, with just one or two plants for each hill.

Bill told me that he harvests his gourds after 140 days, when the vines have shriveled and gone through a couple of frosts to "cure" them. (I don't have frosts in my California gardens, and waited until the vines had thoroughly browned and died back.) Bill leaves on a few inches of the stem and stores the gourds in a cool, dry garage for five or six months.

After the gourds are dry and rattle slightly when shaken, Bill cuts a two-inch entry hole three to five inches up the side of the gourd, and cleans out the seeds. With his pocket knife or a paring knife, he enlarges the bottom of the hole just one-quarter inch so that it is pear shaped. According to Bill, this small modification makes it easier for the martins to carry nesting materials into the houses. No perch is necessary at the entrance hole: The martins don't need one, while aggressive birds such as starlings would happily use a perch to gain easy access to a ready-made home.

With a quarter-inch drill bit, Bill bores a series of holes around the top and bottom of the gourd to provide ventilation and drainage. A small hole at the very top of the gourd is strung with a plastic-coated wire hanger, which can be attached to a gourd rack or to a plastic-coated wire stretched clothesline-style twenty feet above the ground, with gourds spaced a foot or two apart. The hanging hole at the top of the gourd is generally drilled directly above the entry hole.

A METAL COLLAR, MOUNTED
7 to 8 FEET UP THE POLE
PROVIDES PREDATOR PROTECTION

Purple martins are gregarious birds and prefer to nest in colonies, so a collection of at least a dozen gourds works best. Unlike the many species that want to nest far from people, the martins seem to like their colonies located within 100 feet of a human dwelling, which makes martin watching, appreciation, and protection easier.

To clean and preserve my gourds, I washed them with hot, soapy water and soaked them for twenty minutes in a residue-free copper sulfate solution that Louise Chambers of the Purple Martin Conservation Association (PMCA) recommended. I also took her advice and painted the gourds white with an exterior enamel, though I confessed to her that I prefer the natural look of unpainted ones. She said, "I do too, but the martins choose white gourds over the natural. Entrance holes are easier to locate and the reflective white paint keeps interior temperatures much cooler."

Bill commented that gourd houses shouldn't be hung until a month after the first martins have arrived in the spring. Hanging the gourds too early encourages other species to compete for them. He has found, through trial and error, that it is best to situate gourd houses at least forty feet away from tall trees and buildings, so that martins have a clear air corridor for incoming and outgoing flights. Most martin enthusiasts agree that gourd racks or poles supporting the line of gourds should be metal or a sturdy, smooth wood, should be free of any vines or shrubs that could provide access for predators, and should have a collar of tin or aluminum mounted four feet above the ground.

To alert the first arriving martin males that some prime Longwood apartments are ready and waiting, Bill repeatedly plays a cassette recording of a male martin warbling the "dawnsong," a loud, rhythmic melody distinctly different from its daylight singing. He said that the recording is the song of the dominant male in a nesting colony and that it can be heard over thirty square miles from the site! He plays his recording for a few hours each day, beginning at four A.M., until the colony is settled. I can't wait to order my own copy of the tape from the PMCA, but I know that my new neighbors may not be thrilled by the early-morning serenade.

Once the gourds are occupied, they should be checked for nest crashers such as English sparrows or starlings. I asked Bill if the swaying of the hanging gourds caused any problems for the birds. He said that purple martins actually seem to prefer the free-swinging gourd houses to the typical and expensive white wood or metal martin homes that are commercially made, perhaps because the swaying is more closely akin to the martins' ancestral homes in the hollowed-out branches of trees.

Bill explained that autumn cleaning and storing of the gourds is essential for their longevity. He has seen painted gourds that have lasted through thirty nesting seasons! It is amazing that an inexpensive packet of seeds can provide decades of pleasure.

Growing my own birdhouses helped alleviate some of the financial discomfort that sometimes accompanies the purchases of an overzealous gardener. Building birdhouses is another solution for someone who loves the idea of providing homes without spending themselves out of their own. But building a birdhouse is not as simple as it sounds.

All of a bird's nesting requirements must be taken into consideration. What size entry hole fits the species you wish to attract? Are you mounting the house in the proper habitat and at the correct nesting height for the bird? Is the house well ventilated and easy to clean? Is it free of cracks and openings that would allow rain to enter and dampen nesting material and chicks? Once I had figured out purple martin housing, I turned to bluebirds.

Armed with a stack of bluebird information Dave Thompson sent to me, I waded through a sampling of thirty-six-and-a-half years of field work done by Warren Lauder, respected bluebird conservationist. After reviewing his simple, logical bluebird house plans, Jeff and I decided to devote some time to building a few. But before we started building, we had a question about the plans, which inexplicably called for the pine-roofed house to be finished with a dark asphalt shingle.

Mr. Lauder is very modest about his tireless years of bluebird work, but was outspoken and confident about his perfected house plans. He explained that bluebirds are secondary cavity nesters—they occupy cavities, but are incapable of creating them. They were displaced and threatened when introduced English sparrows and starlings took over their tree-hole nesting sites. In 1960 he became aware of the plight of the bluebirds when he found out how drastically their numbers had declined. He read everything he could find about bluebirds and the new nest boxes being built to save them.

His detective work discovered all the flaws in the boxes that were being constructed. Rain entered the houses and chilled the nestlings, the heat of summer sometimes killed the embryos in the second and third clutches (nests of eggs), and the chicks often died from heat and parasites. Working with a friend from the Engineering Department at DuPont Company, he re-engineered the traditional nesting box to create the "only bluebird house with a positive air exchange when it is needed."

He explained to me how his nesting box differs from others. His houses are built of white pine that has not been painted or finished. He drills vent holes near the top of the box and a $^3/_8$-inch vent hole in the floor, and cuts each corner of the floor at a 45-degree angle, which creates holes that further increase air circulation.

The "Bluebird Man" also redesigned the front panel, lowering it slightly to provide a long slot vent. He hinges the front panel with one nail on each side of the box, making it easy to lift up for monitoring nestlings and cleaning the boxes between clutches.

While most bluebird plans call for a roof overhang of only one to two inches, the roof of Mr. Lauder's bluebird house overhangs $1^3/_8$ inches on the sides, $^1/_2$ inch on the back, and $3^3/_4$ inches on the front. The wider overhang shields the nestlings from extreme heat and rain and the probing paws of predators. Inside, deep cross-hatched cuts in the front panel enable fledglings to climb to the entrance hole. (Some modifications in the entrance-hole size will be needed to accommodate the western and mountain species.) His houses are mounted five to seven feet high on metal fence posts or pipes and placed 100 yards apart.

"My houses have the traditional $^3/_4$-inch-thick pine roof, but they are topped off with a dark asphalt shingle," he explained. At this point, I interrupted Mr. Lauder and told him that the shingle was the one thing that hadn't made sense to me. He chuckled and said, "The heat buildup on the roof (because of the dark-colored shingle) causes the air to rise and exit through the vent holes, creating a vacuum which pulls the outside air through the cut-off floor panel. That is what I mean by a positive air exchange. It keeps the temperatures tolerable for the chicks!"

The "Bluebird Man" didn't have to give me any more explanations. His successful houses have dramatically and significantly increased the population of bluebirds wherever they are used. We hope to add our gardens to his log of bluebird success stories.

My long list of garden "necessities" has barely had an impact on our clothing, food, and survival budget, thanks to the help of Bill Haldeman and Warren Lauder. Because of them, I'll have food for my body and food for my soul every spring and summer morning, when the martin warbles a dawnsong and the bluebirds sing what Thoreau described as their "soft, flowing, curling warble."

Despite what Imelda Marcos feels, I believe that a hundred pairs of shoes aren't worth one of those mornings.

In the Company
of Owls

"A wide-eyed owl sat in an oak,
The more he saw, the less he spoke.
The less he spoke, the more he heard,
A valuable lesson from a wise old bird."

—ANONYMOUS

Barn owls - My future weed patrol

"Another day with nothing accomplished," I said aloud as I surveyed the raised plots of ground waiting to be planted. I spent more time gabbing and swapping plant tales than I did working. Daylight hours in my village gardens always bring dozens of visitors with hundreds of questions and stories to tell. I knew that if I wanted to finish my projects in time for our annual Children's Faerie Festival I would have to do something drastic. Desperate for time alone in the gardens, I turned my schedule topsy-turvy and began my "work day" when most people are getting ready for bed.

I settled easily into the routine of the night shift. No need for gooey sunscreen or floppy straw hats. No worry about the spaghetti sauce stain that formed a perfect outline of Florida on my shirt. And no comments about the large spoon that serves as my favorite gardening tool. I set to work eagerly each evening, and accomplished more in three hours than I could in a full day of work.

Although humans weren't present in the slow hours between dinner and midnight, I had plenty of company. The faint, chicklike murmurations of western toads sounded from half a dozen damp, hidden havens. A boisterous chorus of frogs mingled with the lilting songs of the mockingbird who claimed the blackberry hedge.

Another familiar sound, but not as welcome, came from under the drooping elderberry bush. Mice, the bane of my gardens, skittered and squeaked through the carpet of dry leaves as they made forays to the bird feeder and plundered my compost bin for hapless worms.

As I tucked the last alpine strawberry into a border late one night, an eerie, unidentifiable string of staccato clicks echoed throughout the garden. I turned on the flashlight and shined it into the trees and hedge and underneath the rickety wooden porch. Nothing unusual appeared, so I switched off the light and began raking the granite pathways.

The strange clicking resumed and I glanced up as something large flew silently past. High above the gardens four common barn owls (*Tyto alba*), their buff-colored bodies glowing in the light of a lone street lamp, wheeled and tumbled through the starlit sky. I

dropped my rake and stretched out on the ground to watch and listen as they circled and clicked in unison.

Common barn owls are far from "common" today. Though they are found nearly worldwide, their numbers have dwindled. More than a half dozen states in the United States list them as an "endangered species," and in Canada they are designated as a "vulnerable" species. Scientists attribute their decline to a loss of habitat, prey, and nesting sites, and the use of pesticides and rodent poisons that cause reproductive failure and death.

I was thrilled to see the winged foursome cavorting above my mouse-plagued gardens. Barn owls are famous for their "mousing" skills which are better than the most accomplished of house cats. The Owl Rehabilitation Research Foundation in Ontario, Canada, credits barn owls with consuming twice as much food for their weight as any other owl, making them one of the world's most beneficial birds. Scientists have observed nesting pairs delivering more than two dozen rodents (moles, gophers, rats, voles, and mice) a night to their hungry youngsters. Since fledglings don't leave their nest for nearly two months, the parents may capture as many as 1,440 rodents per clutch.

As the quartet passed above me, one of them looked down. I could see the white, heart-shaped face and dark eyes that earned this bird the nickname of "monkey-faced owl." These wide-eyed birds have rod-rich retinas (the part of the eyeball that is sensitive to light and receives images) that enable them to see better in the dark than humans can in bright daylight. Couple those all-seeing eyes with an acute sense of hearing, and it is clear how barn owls can track prey unerringly on the blackest of nights.

A piercing, raspy hiss that sounded like the steam escaping from my Mom's old pressure cooker filled the night sky. The attenuated "whisssssss" is the vocalization most common to these birds. Though vociferous when mating and communicating with their kind, they are the stealthiest of hunters. Mother Nature outfitted them with lightweight, loosely plumaged bodies and comblike feathers with long, fine fringes that act as sound mufflers. Their magnificent forty-four-inch wingspan enables the birds to lift off noiselessly (no extra flapping for them) and glide effortlessly, the better to locate and surprise prey.

I heard the gate latch snap as my husband walked from the barnyard. He sat with me and together we witnessed the owls' sky dance and heard their raucous conversation. We watched them until they broke their formation and flew out of sight. I pushed my tired body up from the cold ground, picked up the rake, and called it a "day."

Before I climbed into bed that night, I searched the house for a copy of David Mas Masumoto's book *Epitaph for a Peach*. David is a second-generation California farmer who writes eloquently and gardens organically. I recalled reading a passage in his book that mentioned a serious rodent problem in his vineyards. David hoped that

• Attracting Barn Owls •

- Provide an enclosed nesting box at least 18 by 18 by 24 inches with at least a 9- by 9-inch entry hole.
- Construct the nesting box of marine-grade plywood or use untreated, recycled wood.
- Extend a lipped wood "porch" under the doorway— for an exercise yard for owlets testing their wings.
- Mount the box 15- to 30-feet above the ground or inside an old barn or outbuilding in a quiet area free from human disturbance (roads, pathways, work areas, etc.).
- Squirrels, snakes, opossums, and raccoons will eat the owls' eggs and prey on young owlets. Provide a 15-inch metal collar or sleeve on posts or trees, or place your nesting box on top of a metal pipe.
- Place the box in an area protected from extreme heat.
- Make sure there is an unobstructed flight path to the box.
- Build boxes this winter and install them outdoors in appropriate areas. Barn owls (and other owls) seek shelter in them during severe weather long before they utilize them as nests.

if he built nesting boxes and placed them on his property he could tempt the owls back onto his land. Barn owls are not nest builders. They are cavity nesters who utilize caves, tree holes, barns (and any buildings with accessible openings), and wooden nesting boxes. I wanted to learn more about David's boxes but couldn't locate the missing book, so I decided to telephone him in the morning.

"Oh, the owl boxes worked out wonderfully," David said when I asked him about his experiment. "We used old barn wood to make them. They are two feet by three feet with a one-foot-square entrance opening. I mounted the boxes fifteen feet high overlooking the eastern view of the vineyard. The east side of the building is not as hot as the other side, and our family can sit on the porch and watch the barn owls raise their families. In the early spring we are serenaded by the sounds of the chicks and entertained by their parents as they deliver food to them. The kids can actually watch the owls swooping down and catching the rodents! It is all part of the balance of nature and this is the way my kids learn about it. It is just great for my family to be a part of this." Before we said our good-byes, David mentioned that the owls have solved his rodent problems and that he hasn't seen any more signs of damage. He added, "A few of the large, fresh-market grape growers have now put up owl boxes. They seem to work and I think other farmers are going to give them a try."

I'd like to give them a try, too. Building a couple of nesting boxes would be a great project for the end of the year. Some of my bird books mention that barn owls begin searching for nesting sites in early January and that in cold weather they use them as a roosting area. I'd love to start this year with a barn owl, or better yet a family of owls, in my gardens. Come spring, when I'm working alone by moonlight, perhaps they'll do an encore performance for me.

A Charm of Finches

"A very interesting feature of our bird-songs is the wing-song, or song of ecstasy. It is not the gift of many of our birds. Indeed, less than a dozen species are known to me as ever singing on the wing."
—JOHN BURROUGHS, *Ways of Nature*, 1905

It stormed wildly last night, and our small Maine cottage shuddered with each thunderous wave that slammed against the rock ledges. Outside my window I could see a skeleton moon peeking from behind the thick clouds that scudded toward Pemaquid Point. The slender white limbs of the birches scratched against the window screen, and the fir and spruce trees seemed to wail in response to each powerful gust of wind.

I grabbed my bathrobe and flashlight and slipped quietly down the hall and out the front door. I was worried about the welfare of my "garden" of container plants, which crowds along the edge of the porch and down the steps to the blueberry patch. And I was anxious about the goldfinch nest woven into the outermost branches of an oak tree.

It took only a few minutes for me to nudge and shove the heavy containers snugly against the sheltering wall; then I turned my attention to the thrashing oak on the bluff. My flashlight illuminated the compact nest, which bobbed and pitched like the boats moored in Little Harbor. Just a few hours earlier I had watched as the brilliant yellow male goldfinch made repeated trips to the nest with food for the soberly clad and perfectly camouflaged female, who never left her eggs untended.

The goldfinches (*Carduelis tristis*) are the first birds to discover our feeder each spring. Although some people complain that they are greedy eaters, we feel thankful for their presence and are happy to replenish their food supply, sometimes twice a day. Our reward is that they entertain us from sunrise to darkness with their cheerful songs, lilting flight, and comedic jostling at the feeder.

Early-twentieth-century author Thornton W. Burgess called the goldfinches "Chicorees", for their signature "per-chic-oree" song, which they perform on the wing—a rarity in the bird world. My Gram McKinstry called them "wild canaries", and Grandma Clarke referred to them as "thistle birds." I like the old fashioned name "farmer's friend" bestowed on them by agrarians who recognized their value.

Goldfinches are voracious eaters and are credited with consuming thousands of pounds of noxious weed seeds annually. Ragweed,

thistle, shepherd's purse, and dan-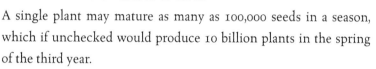
delion are their major food
preferences, with a small helping of aphids and caterpillars
as a side dish. A 1917 study
by Dr. Sylvester Judd of the
U.S. Biological Survey noted
that certain garden weeds produce an incredible number of seeds.
A single plant may mature as many as 100,000 seeds in a season,
which if unchecked would produce 10 billion plants in the spring
of the third year.

Although always first in the food line, the goldfinches are the
last to mate and nest in our garden. Just as the chickadees, phoebes,
jays, and wrens are fledging their young, the goldfinches begin setting up housekeeping. Their timetable is dictated by Mother Nature,
who provides them with an abundance of seed to feed themselves
and their young and a supply of the thistledown used to upholster
the inside of the nest. During a nearly week-long construction
period, the female selects fibers and secretly weaves and lines a shallow cup that will one day cradle four to six pale blue eggs.

The winds accelerated and the row of rocking chairs on the
porch pitched back and forth as though peopled with invisible
onlookers. As my flashlight dimmed, I gave up hope of seeing
whether the female was still on the nest, and headed back indoors.

I slept fitfully that night and awoke just as the sun flushed the
sky and sea a deep rose. In the garden, only a scattering of branches
and twigs, a fallen bird feeder, and a meadow of wind-bent grasses
told the tale of last night's storm. I filled the wire-tubed feeder with

• Bring on the Birds •

- Plant a pot or a plot for the goldfinches and resist the urge to deadhead or tidy this area.

- Some of their seed-bearing favorites are bachelor's buttons, coneflower, coreopsis, catmint (*Nepeta*), milkweeds (*Asclepias*), cosmos, sunflowers, millet, evening primrose, love-in-a-mist, black-eyed Susan (*Rudbeckia*), Tithonia, pincushion flower (*Scabiosa*), and goldenrod.

- Fill a basket, mesh onion bag, or small box with narrow grasses, fine strips of bark, thistle, milkweed, and cattail down. Hang or place it in a sheltered area protected from rain and cats (I put mine in a hedge). Your offerings will be used in the construction and lining of nests.

- Goldfinches happily visit feeders filled with black-oil sunflower seeds, or mesh socks or special feeders stocked with niger, a tiny, black thistle seed.

black-oil sunflower seeds and slipped it into place, then walked beneath the branches of the oak to search for the remnants of the nest, or worse yet, nestlings in distress.

Behind me a flurry of finches fluttered from the trees like autumn aspen leaves and landed on the sides of the wire feeder. As the birds plucked and hulled the seeds, they somehow managed to keep up an endless stream of twittering, which naturalist Gene Stratton Porter called their "See me?" song. The Middle English appellation for a group of finches is a "charm"—from the word *cherm*, which means a blended singing, or incantation, of birds. My sprightly charm of goldfinches lived up to their name and serenaded me until an aggressive and noisy blue jay swooped down to displace them.

The "per-chic-oree" call of a male sounded, and I looked up just in time to see him land on the edge of the nest, which was, remarkably, still intact. He dipped into the cup and stuffed some food into the raised bill of the female, then flew off toward the woods.

After the male left the oak, I moved our extension ladder from the side of the cottage to the tree trunk, climbed a few rungs, and peered down into the nest. The trim little female looked at me but didn't move. Despite raging winds and pelting rains, the goldfinch nest was as fit as the day it was built. I didn't want to alarm the bird and paused only long enough to note that she had woven her thistledown lining around a cluster of ripening acorns. My worries about the future family cradled in the oak branches were at rest for the day.

At the rate the birds are emptying our feeder, I will soon need to make another trip into town for seed. But, until then, I am content to sit and visit with the charm of finches that fills my garden with golden light and a tapestry of sounds.

Something to Crow About

These intelligent birds know a thing or two about family values.

After years of estrangement, the crows and I are finally reestablishing our old relationship. It is slow going and a bit tenuous, but they seem to be tolerating my presence in "their" garden, the one I tend for them. For years, the big, black birds coexisted peacefully with me. Each morning when I ventured outdoors to fill the feeders and bird baths, the vociferous family of crows who roosted in the neighboring forest hushed their conversations and flew silently through the yard. They alighted on the limbs of an old cedar tree and watched as I moved along the pathway toward their feeding tray on top of the wood box. As soon as I filled the tray with crowbread and suet and walked away, they swooped down for their hearty breakfast.

After their meal the crows dispersed throughout the gardens and mined the ground for an assortment of grubs and bugs. I enjoyed their company and loved the fact that they helped me with a sometimes disagreeable garden chore.

I remember reading in William Bartram's famed 1791 book *Bartram's Travels and Other Writings* about a tame crow named Tom, who not only ate his share of bugs but also labored alongside the author in the garden. When Bartram weeded, Tom, who like all crows was a great imitator, set to work pulling weeds, too.

I always prided myself on understanding the American crows (*Corvus brachyrhynchos*, Greek for "short snout") who shared their garden with me. While I raked the wide drive under their treetop perches, I listened attentively to their medley of communicative calls and observed their most intimate family interactions. I recognized the soft murmurings between a mated pair, the harsh squawks of an overworked parent, and the warning—clacking sounds, like castanets, that signaled a male's dominion over his mate and territory.

Perhaps my favorite behavior was when they gathered in a mob, shrieked their wild, hoarse caws, and harassed their enemy, the Great Horned owl, who had the audacity to fly into their territory. They surrounded this night-hunting bird who often plucked the sleeping crows from their roosts like grapes from a vine, and tormented him until he took off for a more peaceful resting place.

• Corn-fusion •

To be truthful I never expected this old-fashioned foolery to work with the intelligent crows, but it did. I learned the trick from an Alabama gardener, then read another version of it in a book of North Carolina folklore.

To keep crows from bothering your garden plot or young corn, pound poles around the perimeter of the bed and tie a long piece of twine or string to one post. Then string the twine back and forth and crossways (it will look almost like a spider web). The crows think this is a net or a trap and will avoid flying into the area.

Many Americans fear that the traditional nuclear family is a thing of the past, but in the world of crows the institution is as strong as ever. Barring unforeseen catastrophes, crows mate for life, and often many generations of a family live together and assist in chores as varied as acting as sentinels, building nests, locating food sources, and raising young. Adaptable, omnivorous, sociable, and one of the Einsteins of the bird world, the intelligent crows flourish in an environment that is sometimes most inhospitable to them and their brethren.

From my studio, I had a bird's-eye view of the crows as they built a large, treetop nest of twigs, grass, and fibrous strips of bark. When the female settled down to brood her clutch of freckled blue eggs, the family visited her regularly and kept her well fed and entertained. She occasionally left her duties for a few minutes of recreation, but one or two of the crows always stood guard beside the nest until she returned.

Although I was accustomed to the crows' morning hunting expeditions, I was shocked by the number of creatures they gathered for the nestlings. The line of birds delivering food made the sky around the treetop look like Chicago's O'Hare Airport on a busy day, and the commotion from the youngsters sounded like a group of teenagers at a slumber party.

The crows seemed to know instinctively that earwigs, snails, and slugs are plentiful in the container garden; that cabbage moth caterpillars hide under the nasturtium leaves; and, sadly, that I had just enriched my soil with buckets of worms. They gathered quickly and returned often, and my borders became as clean as a freshly vacuumed carpet.

A recent New York study estimated that corn comprises only one percent of a nesting crow's diet, and that much of the corn consumed is gleaned after the harvest. A typical family of crows may consume more than 40,000 agricultural pests during their nesting season. Admittedly, the birds do sometimes take nestlings and eggs of other birds, which makes them unpopular with many people, but their predations do not significantly impact the songbird population.

Family obligations didn't end when the scruffy-looking youngsters outside my window fledged. For months afterward, they shadowed the adults and begged for food. From my vantage point, it

• Crowbread •

Crows will love to feast on this healthy snack. Others birds will savor it, too.

2 whole eggs
2 cups flour
3 teaspoons baking powder
2 cups yellow cornmeal
1 ½ cups milk
4 strips of bacon, cooked and crumbled
1 tablespoon bacon drippings

Beat eggs and crushed shells together. Sift flour and baking powder and add cornmeal, milk, crumbled bacon, and bacon drippings. Mix together and pour into a greased and heated 8-inch iron pan and bake in a 400°F oven for a half hour. Cool and cut into squares. (I feed them one or two squares daily and keep the remainder refrigerated.)

appeared that the demanding young crows were pesky and irksome, but the older birds fed them patiently and taught them the tricks of survival and the etiquette of coexisting in a large family.

My peaceful morning ritual of feeding the birds continued until the day I fell from grace. I heard a clamor of crows in the pathway and discovered an injured youngster missing a leg. I gently wrapped her in a blanket and tucked her into a cardboard box for a trip to our veterinarian. As I walked to the car, the boisterous family of crows flew from tree to tree along the trail and reprimanded me loudly.

I returned home that afternoon without the youngster, who could never survive in the wild. Our local wildlife rehabilitation team decided to tame the glossy black bird and include it in a series of educational programs for schoolchildren. The crows were in the garden when I got out of my car, and I heard the hoarse, insistent cawing they used to chastise trespassing Great Horned owls. The birds descended and scolded insistently, and I realized that they were mobbing me—not only the woman who fed them, but also the person who had "kidnapped" one of their family.

Whoever claimed that elephants are endowed with the best memories was wrong. Crows never forget anything, and they have the ability to recognize an individual offender (in this case, me) in a crowd. They routinely mobbed me—even as I filled their feeder—whenever I ventured outdoors. Now, new generations outnumber the old, and I can again work peacefully in the gardens. The crows visit their feeding tray, harvest unwanted pests from the borders, and entertain me with their complex family interactions. Everything would be perfect if only I could teach them to help me with the weeding.

Jay Walking

"The blue jays evidently notify each other of the presence of an intruder, and will sometimes make a great chattering about it, and so communicate the alarm to other birds and to beasts."
—HENRY DAVID THOREAU'S JOURNAL, DECEMBER 31, 1850

Jays are the rascals of my gardens—hooligans, tattletales, and belligerent thieves; but they are also beautiful, intelligent, and bold. I have a special place in my garden and in my heart for these brilliantly colored birds, but it wasn't always so.

"You have to learn to take the bad along with the good," my Grandma Clarke said to me when I was a child. She peered at me through her thick spectacles, smiled, and said, "You'll find that's true of all of life." My Grandma was referring to a jaunty, blue scrub jay (*Aphelocoma coerulescens*) that claimed her handkerchief-sized garden as its territory. Every time I poked my head outside the screen door, the jay scolded me, swooped down, and then shadowed me threateningly as I ran across the yard.

As I shuffled through the grass beneath Grandma's avocado tree late one afternoon, a small but clamorous outcry sounded from the branches above me. I stopped, and the noise stopped. I shuffled again, and the noise resumed. I looked into the shadows of the tree but couldn't see anything, so I leaned my Grandpa's ladder against the trunk and climbed into the branches.

My movements up the tree triggered another round of noise, and I inched my way toward the sounds coming from a loosely constructed nest of twigs. I had only a moment to peer into the nest of naked baby jays before the parent returned. I think we shocked each other equally, but I was the one who fell backward from the tree and onto the grass.

When I caught my breath, I moved the ladder to a safe distance and climbed onto the top step to watch the nest and the hunting forays of the parent. From my perch I could see the jay picking through our neighbor's newly planted vegetable patch and watched as it plucked something from a leaf. The jay flew to a low branch of an apricot tree, and then wove its way back to the vicinity of the nest by a circuitous route, from bush to tree to bush. It sat for a moment with a fat caterpillar in its mouth, checked one last time to make sure that no predators were near, and flew across the last few feet into the treetop.

Between feedings, the young were silent, but as the parent swept through the branches and landed on a nearby twig, the four youngsters squeaked as loudly as my Grandma's old screen door. I sat awhile and watched as the jay made numerous trips into the neighbor's garden. Each time, it returned with a beakful of caterpillar, wasp (how did it capture it without being stung?), fly, beetle, or some tough-looking grasshopper.

By nightfall I had a greater appreciation for the jay both as a protective parent and provider. I learned that, contrary to what my Grandpa had told me many times, jays are not just nest robbers; they're great garden pest controllers. I was no longer afraid of them, and looked forward to welcoming them into my own garden someday.

Eight species of jays are year-round residents of the United States. Gray-breasted jays occur in southern Arizona and Texas, green and brown jays are tropical species found at the southern tip of Texas, and Pinyon jays dwell in the mountains and high plateaus of the west. The fluffy gray jay's range extends from the southern Rockies and Pacific Northwest through Canada and into Alaska. The territory of the brilliantly colored blue jays, usually thought of as an Eastern or Midwestern species, is expanding slowly westward. The scrub and Steller's jays are dwellers of the West, with a tiny, isolated subspecies of scrub jay found in Florida.

My childhood dream was realized and now both scrub and Steller's jays make their homes year-round in my California garden, or should I say that they allow us to use their garden? They seem to be the dominant species of bird residing here, and I have to admit that nothing escapes their attention. In past years some of the jays were nearly tame, and one particular Steller's seemed smitten with my son, Noah. Sometimes when Noah walked his big white shepherd, Una, the Steller's followed behind them, flying from branch to branch as they circled the neighborhood.

For the past five nesting seasons, a pair of scrub jays chose the trellis and Cape honeysuckle outside our guestroom window for their "hidden" nest site. I can open the window a few inches and peer down into a poorly constructed nest of twigs that looks like it couldn't hold an egg, let alone the typical four to six nestlings. On a beam over our porch, the Steller's jays annually build an unwieldy nest of mud, twigs, dog hair, shredded paper, and anything else they can find. The large, fifteen-inch nest looks like something that a recycling crew assembled after a long day of work.

On warm spring days, I like to sit near the nests to watch the feeding activity and listen to the jays' soft, chuckling conversations. Most people are used to the strident screams of indignant jays mobbing hawks, owls, or cats, but few notice their repertoire of less-noisy calls and their ability to mimic other species of birds. As the jays converse with each other and their young, they keep a constant watch for predators and somehow maintain a feeding schedule that would tire an Olympic gold medalist.

Last spring I kept a tally of one scrub jay's deliveries to her nest over a three-hour period. Although I couldn't see the species of the insects the jay crammed into the gaping mouths of her young, I counted an average of about nineteen feedings per hour. Since jays forage in the garden from sunrise to sunset, I estimated that they feed their young nearly 200 insects daily.

The jays' young need a constant, high-protein diet of insects, but adult jays are opportunistic and feed on a diverse array of food. Their diets include insects, nestlings, grains, seeds, nuts, berries, acorns, and—to my amazement—frogs, lizards, and an occasional snail.

I once described the jay's eating habits as gluttonous, since they do seem to overindulge whenever they find an abundance of food.

I've learned differently. I changed my mind after I watched a jay pick up eleven sunflower seeds in rapid-fire succession before he left the feeder and flew to the top shelf of my potting bench. Being curious, I climbed to a vantage point on the upper trail and watched as the jay slowly poked most of the sunflower seeds into my soil-filled peat pots. "Aha!" I exclaimed, as the jay exposed a new facet of his personality. Throughout the gardens, sunflowers and young oak trees were popping up in unexpected places. The hoarding instincts of the jays were paying off, as they cached their seeds and acorns throughout my gardens and containers and promptly forgot where they were hidden. Mother Nature's version of Johnny Appleseed—and I had accused them of gluttony!

Last spring the small, copper-roofed birdhouse outside my kitchen was filled with a succession of chickadee families. Normally, I never interfere with the natural occurrences of nature. It is difficult, but I try to let nature balance itself without my intrusion. The battle between the chickadees and jays is one exception. I am attuned to the alarm calls of the chickadee parents and found myself responding to their high-pitched distress calls about a dozen times a day.

I ran outdoors flapping my apron wildly, admonishing the jays to abandon their raids on the chickadee nestlings. Sometimes I won, but a few times the jays were victorious. The chickadees acted agitated and searched for their young for a few minutes, then returned to their normal routines.

I think I was more upset than the chickadees and felt bad for hours. I was responding in a very unscientific manner to something that is a natural occurrence. That evening, as the chickadees retired and the jays settled down in their nests, I thought back to

• Welcome Jays to Your Garden •

1. Keep one large birdbath filled with fresh water for the jays' bathing and drinking needs.

2. Establish a feeding station in a clearing or meadow, away from bushes that could harbor cats or other predators.

3. Jays love peanuts. Indulge them occasionally with a supply of unsalted raw peanuts placed on a platform feeder, beam, or ledge. Locate it near a window so you can enjoy their antics.

4. Plant a portion of your garden with native oaks, pines, and berries to furnish the jays with some of their favorite foods.

5. Fill a wire basket or onion bag with hair, shredded paper, string, grass fibers, tiny twigs, and pine needles, and hang it from a tree. The jays will peruse your offerings and tug out anything suitable for their nesting needs.

6. Avoid the use of pesticides, especially during the nesting season when the jays depend on a supply of insects.

my childhood and the tried and true wisdom my Grandma Clarke dispensed as readily as her cookies: "You have to accept the bad along with the good." I looked across the garden at the dozens of young oak trees and sunflowers planted by the jays. Grandma was right, as usual.

Scarecrows Are for the Birds

Shoo-hoo, shoo-hoo!
Away, birds, away,
Take a corn, and leave a corn,
And come
No more today!

—OLD SCARECROW SONG

1812 FARM
BRISTOL MILLS, MAINE

I saw him in a field north of Kokomo, Indiana, languidly bowing and waving his arms above a fluttering stand of knee-high July corn. Nothing in my suburban, Southern California background had quite prepared me for such a gentleman.

We pulled the car over to the edge of the road and took a dozen photographs. My husband shook his head, amazed that a simple scarecrow could stir up such excitement in me. I tried to explain that aside from my childhood love for the brainless straw man in L. Frank Baum's *Wizard of Oz* stories, and Ray Bolger dancing down the yellow-brick road with Dorothy, I had never seen a real, in-the-straw, over-the-field scarecrow. From that day on I was hooked, and scanned every farm and garden as we drove across the country. Whenever I spied a particularly special specimen, we stopped the car so I could sketch and photograph the winsome character. Whenever I could talk with the farmers or gardeners who had built them, I found to my amazement that the old adage about people resembling their dogs was all wrong. They look like their scarecrows, or vice versa.

My straw-man reveries increased and the search for bits and pieces of scarecrow history began to fill my file drawers. At first I believed that some ingenious, thrifty Yankee, frustrated by the visits of hungry crows, had outfitted stick-like forms with some of his worn-out clothing. But, after decades of research, I've found that my ethnocentric musings were off by a few continents and about 3,000 years. Scarecrows, though of humble, tattered origin, have a lineage that outdistances the oldest of royal families. They have been our garden companions and guardians since we first began practicing and depending upon the art and science of agriculture.

The first books I perused in my search for information focused on human "bird-shooers," who watched over seedlings and crops and protected them from the onslaught of hungry birds. The earliest record of living scarecrows comes from hieroglyphics and murals in the ancient Egyptian tomb of Nakht. There on the walls are depictions of men and women waving their arms and hurling sticks to frighten ducks, geese, quail, and gulls away from the ripening wheat. Early written works mention that the hardworking human scarecrows had to use innumerable methods to outsmart marauding birds, especially the intelligent crow. They used wooden clappers, bells, clacking sticks, and metal objects clanged together to frighten off unwelcome snackers. Usually, the bird-scarers were young boys not yet big and strong enough for other jobs, or older men who could no longer work in a skilled trade or on the farm. They had a number of different names in various cultures, such as "Tattie-doolies", "Jack O'Kent", "Tattie-bogles", "Mawkin", "Hodmadod", "Mammet", and "Shoy-hoys", which was an actual call used to frighten the birds. Believe it or not, there are still human scarecrows who work in vineyards and orchards across Europe, practicing a trade that spans 3,000 years.

Although I was fascinated by the long, colorful history of the Hodmadods, my true quest was for the origins of folksy, straw-stuffed gentlemen like the one I had spied in the Indiana cornfield. Some historians and folklorists believe that the first constructed scarecrows were used in the fields of wheat cultivated along the fertile banks of the Nile River. Hundreds of years later, the Japanese began their continuing tradition of building tall bamboo *kakashi*, outfitted in rags, twigs, and clattering fish bones, to stand guard over their flooded fields of rice. Greeks and Romans carved lifelike wooden figures to watch over their orchards and vineyards, and this practice was carried by the Romans to the fields of Germany, France, and England, where their crops were protected by lumpy straw men topped with turnip heads.

I wondered whether scarecrows were a part of our early American heritage and placed a call to Del Moore, the reference librarian

• Create Your Own Scarecrow •

Scarecrows are simple to build and are a great way to re-use old clothing that isn't good enough for donations. The more torn and bedraggled the clothing, the better.

What You'll Need:

- A tall wooden post (6–8 feet)
- 2 lath stakes, approx. $^3/_4$-inches thick and $1^1/_2$- to 2-inches wide, one 4-feet long, the other 3-feet long
- Old shirt and pants, gloves and an old hat are optional
- Pillowcase or gourd for a head
- Twine to tie arms, legs, and neck closed
- Acrylic paint for a face
- Straw for stuffing
- Small screws or nails
- Sink the post into the ground at least 18 inches.

for the Colonial Williamsburg Foundation in Virginia. Del said there were scarecrows in the fields around town, and that they would not be included in the landscape if they were not true to the colonial times. He sent me an article that quoted French-born J. Hector St. John Crevecoeur, agriculturist and author, who wrote that "18th-century farmers constructed images of men, made of straw and dressed in cast-off clothing."

Next, I spoke with Jack Larkin, Director of Research and Collections at Old Sturbridge Village in Massachusetts. I asked Jack if

- Attach stakes with screws or nails to form a cross, using the 3-foot section for the crossarm.
- Put shirt on crossarm and tie sleeves closed with twine.
- Put vertical arm of cross into one leg of pants. Tie both legs of pants closed at ankle with twine.
- Tuck shirt into pants, then stuff shirt and pants with straw. Use twine for a belt to secure waist of pants stake.
- Fill gloves with straw and secure to wrists of shirt with twine.
- Paint a face on a pillowcase, then stuff with straw, or use a gourd as a head.
- Tie twine around neck to secure head to body.
- Attach body to post.
- Tie long ribbons, mylar strips, and pieces of metal to arms to clank together and frighten the birds.

he remembered any writings about the use of scarecrows in New England, and he cited a nineteenth-century traveler's account of "figures constructed of a jumble of old clothes, piled up on a stick and wearing a hat." He laughingly suggested that paying a young boy five cents a day to toss stones at marauders would have been regarded as far more effective by New England farmers.

In America's southwest, the Native American Zuni tribe peopled their cornfields with groups of frightening, man-sized figures called "the watchers of the corn sprouts." The watchers were constructed of poles and branches, with faces of rawhide, luxuriant tresses of flowing horsehair, eyes of corn husk rounds, teeth of cornstalk strips, and long, red leather tongues that hung from their gaping mouths. They were mounted in the midst of fields studded with tall cedar poles crowned with prickly leaves and hung with ropes, rags, hide strips, bones, and feathers that blew crazily in the wind.

My favorite story of the Native American use of scarecrows was told to Gilbert L. Wilson in the early twentieth century by Buffalo Bird Woman of the Hidatsa tribe of North Dakota. "We made scarecrows to frighten the crows," she said. "Two sticks were driven into the ground for legs; to these were bound two other sticks, like outstretched arms; on the top was fastened a ball of cast-away skins, or the like, for a head. An old buffalo robe was drawn over the figure and a belt tied around its middle, to make it look like a man." She went on to describe how the women and young girls watched over the crops from a tall platform built in the garden. She said, "We cared for our corn in those days as we would care for a child; for we Indian people loved our gardens, just as a mother loves her children; and we thought that our growing corn liked to hear us sing, just as children like to hear their mother sing to them."

Every region of our planet sports its own adaptations, traditions, and legends about the ancient protector of crops. In Provence, France, the figures are dressed in ribbon-like streamers that blow and wave in the slightest breeze. Germans believe the fatter the scarecrow, the better the harvest. Italian bird-scarers are dressed in the clothes of the most handsome and virile men in the village, and in Portugal they are constructed of the finest straw to ensure a bountiful harvest. In Austria, it is typical for a priest to bless the figures at the onset of each planting season.

Potato Scarecrow to Protect Your Young Lettuce

This scarecrow was called a *lselaad fogel* by the Pennsylvania Germans who used it to scare birds away from their tender, new lettuce crops. Although this funny looking critter will make you laugh, it must look like a threatening hawk to the little birds who normally feast on your seedlings.

- Sink a pointed bamboo stake (point up) into the ground.
- Stick a potato onto the pointed end.
- Stud the potato with turkey or chicken feathers so that it resembles a bird.

Russians build enormous, twelve-foot-tall forms, because the native crows are gigantic.

During World War II, British Home Guard volunteers dressed themselves in tattered clothing and, disguised as scarecrows, stood motionless and on watch for a German invasion from across the Channel. Early settlers in colonial America topped their straw men with heads made of gourds or pumpkins. And, in areas of the United States heavily settled by Germans or Amish, it is not uncommon to find both a male, *bootzamon*, and female, *bootzafrau*, scarecrow standing guard over orchards, strawberry patches, chicken coops, or fields of corn. I love their belief that the pair keep each other company through their long, watchful hours of garden duty.

I have kept the age-old tradition of garden protectors alive by placing a cheerful, gangly, gourd-headed figure smack in the middle of our yard. Three hundred and sixty-four days a year she is just plain Winifred, the lady of the garden. But, on the day of our yearly Children's Faerie Festival, Winifred emerges from her chrysalis of common clothing and becomes our "Faerie Crow," replete with a set of diaphanous white wings. Reigning high above the crowds, she stands draped in flowing robes of cheesecloth sprinkled with a constellation of golden glitter. Atop her head is a circlet of sparkling stars, and in her dangling, gloved hands she holds a magical fairy wand.

Poor old Win. She stands alone the rest of the year above a cluster of large terra-cotta pots, and watches over the sprouting sunflowers and bee-laden hollyhock blossoms. There is only this one day a year when she receives the respect and adulation due a lady of her lineage. Perhaps I should find that picture of my first love, the Hoosier scarecrow, and build one to stand by her side for better or for worse, through all the changing seasons in my gardens.

Outwitted by Squirrels

Despite my best-laid plans,
they have the run of the garden.

I know they have relatively large brains and are extremely intelligent, but is it possible that squirrels can read? Almost every time I jot an entry into my journal about an anticipated harvest or bloom, the squirrels are one step (or a giant leap) ahead of me. I can envision them thumbing through the pages as they note just the right time to pluck the dusky blueberries or nip the fat, green buds of my favorite miniature sunflowers.

This foggy summer morning I grabbed my cup of coffee and headed outdoors to tend my wild birds and bevy of herbs and flowers. I swung open the door and walked into a scene that looked like the aftermath of a tornado. My window boxes and potted plants were torn apart, soil was everywhere, and strewn across the weather-silvered porch was a carpet of gold, orange, and mahogany petals. A tall line of headless, shredded green stalks was all that remained of what should have been a troupe of flamboyant sunflower faces.

The culprit, a tiny eight-inch red squirrel, whose family sports the nicknames of "fairy-diddle" and "chickaree", betrayed himself by nervously flicking his thick bottlebrush tail from behind a terracotta pot. As I approached the porch railing, he clambered onto a tall container and chattered and scolded me for interrupting his breakfast celebration. I watched as he nibbled at a paw full of tender, newly formed sunflower seeds. "Out!" I shrieked as I stamped my foot in anger. The audacious rodent (from the Latin verb *rodere*, "to gnaw") ran across the porch toward me, scurried between my feet, and dived beneath the rungs of my favorite rocking chair.

In Maine, I am blessed with both the raucous red (*Tamiasciurus*, which means "hoarder") and twelve-inch gray (*Sciurus*, which is a corruption of the Greek word *skiouros* or "shadow-tail") squirrels. When times are tough and food supplies are scarce, the gray squirrels sometimes emigrate en masse to new surroundings. John James Audubon and John Bachman in *The Quadrupeds of North America* wrote, "Onward they come, devouring on their way every thing that is suited to their taste." One historical document written by Ernest Thompson Seton, and verified by numerous scientists, describes an 1842 emigration of gray squirrels from Wisconsin, which numbered in the millions and lasted for four weeks. Red squirrels are also travelers and sometimes swim across broad expanses of water in search of richer feeding grounds. This morning, as I looked at my devastated sunflowers and disheveled containers, I felt as though most of the footloose squirrels on the East Coast had somehow ended up in my island garden.

Although the pugnacious squirrels are a constant challenge (i.e., problem), I often stop my work and watch in awe as they leap nimbly along their treetop highways. Squirrels use their tail like a counterbalance, turning and dipping in response to swaying branches and precarious perches.

When they reach a wide clearing, which zoologists poetically term an "air bridge," they push off with their strong hind legs, flatten their bodies glider-style, and employ their aileron tails for banking, turning, and slowing their speed. The squirrels are sometimes able to kite their way twenty feet to a target, which unfortunately is often a "squirrel-proof" birdfeeder.

During rainstorms, their furry tail, which curls up their back like a large question mark, acts as an umbrella, shielding them from wind and rain. Biologist Adrian Forsyth, Ph.D., believes that they also utilize their tail as a sun screen, raising it over their body like a parasol. In winter, when their tails are most luxuriant, they nestle inside the snug, encircling coil of fur only to emerge when the weather warms enough for foraging.

Our gray and red squirrels build their dreys (a Middle English word for their nests) of moss, twigs, leaves, and bark high in the

· Helpful Hints ·

To protect fruit trees, the Humane Society of the United States recommends that you trim any branches less than six feet from the ground and wrap a two-foot-wide band of sheet metal around the trunk about six feet above the ground. Dwarf trees can be covered with squirrel-proof netting. Remember that if there are nearby trees, roofs, or other access points, the squirrels can and will make the leap.

- **Spray your flowers and shrubs** with a potent concoction of fermented salmon; give your plants another spray after rainfall.

- **Protect your tiny bulbs** (grape hyacinths, snowdrops, miniature jonquils, etc.) by tucking them inside a wire basket and surrounding them with crushed gravel, through which the squirrels dislike digging.

- **Place bird feeders on a pole** in a clearing and use a baffle to stop squirrel raids (you may need to use a baffle above as well as below the feeder).

Caution: Female squirrels scout for safe nesting areas and will seize any opportunity (a rotten board, open vent, or hole) to enter an attic. Once inside, they may gnaw through insulation, wall boards, and wiring. If squirrels are present in your yard, check your attic for entry points and cover holes with metal or mesh before the squirrels can enter.

• For Squirrel Lovers •

If you actually want to attract squirrels, purchase a special squirrel feeder, fill it with peanuts, and situate it far from other feeders in your garden. I enjoy watching the squirrels figure out how to operate the lift-top feeders.

Squirrels utilize nesting boxes for raising their families and for shelter. Purchase one with at least a three-inch entry hole located on one side of the box to provide easy access.

branches of trees or inside hollow boles or man-made nesting boxes. After a frenzied, trunk-spiraling courtship chase, the squirrels mate, the male is banished, and thirty-eight to forty-five days later a litter of up to eight young is born.

Our garden is relatively quiet during the first two months of the youngsters' lives, but after weaning we watch as the diligent females deliver berries, seeds, eggs (I don't like that, but it is nature's way), and huge mushrooms to their hungry offspring. Soon, the female forgets her motherly instincts and drives the young out of her territory.

Throughout our granite-ledged gardens, gray squirrels excavate innumerable shallow holes for their stashes of food. Their territorial cousins, the red squirrels, tend only one or two large caches, which they constantly stock with provisions. Both species turn to their reserves whenever there is severe weather or a shortage of food. They literally sniff out their supplies and can locate them under several feet of snow. Thanks to unused and forgotten storehouses, squirrels are responsible for the dispersal of beneficial hypogeous (subterranean) fungi and seeds and for planting hundreds of thousands of trees each year.

This afternoon I'll drop by Wind Song Nursery and try to replace my damaged plants with a new crop of sunflowers. I'll spray the buds with a potent concoction of Coast of Maine Fermented Salmon, the same mixture that finally stopped the chipmunks from devouring my cherry tomatoes. After dinner tonight, I plan to order a squirrel peanut feeder (guaranteed to keep them busy) and a few nesting boxes, which we are going to mount far from our porch garden and the bird feeder. Perhaps these offerings and deterrents will keep the squirrels happy, but just for added insurance, I think I'll hide my journal inside the old captain's chest after I finish the day's entries.

On the Matter of Moles

*Gardeners, please heed my
gentle plea for tolerance.*

Late one mild winter night, when the fingernails-on-chalkboard shrieks of a barn owl roused me from sleep and lured me outdoors, I rediscovered an old friend. I stumbled upon him quite by accident as I followed the narrow beam of my flashlight along the garden pathway near the bulb-starred woodland border.

My light caught the silvery glimmer of a snail trail, and as I crouched to search for the culprit before he did any damage, a knot of snowdrops and grape hyacinths shuddered, shifted, then tilted awkwardly onto its side. I watched as the soil lifted and folded like the pleats in a skirt, then rose slowly into a meandering ridge of ground. Within minutes, the soft line of earth ran about three feet through the border, and my bulbs looked like jetsam strewn along a storm-tossed coastline.

"Diggory, diggory delvet, a little old man in black velvet," I said aloud as I remembered the old rhyme from my childhood. The broad-footed mole, *Scapanus* (from the Greek word for spade) *latimanus* (for wide hand), had reclaimed the rich hunting grounds of my garden and now swam a perfect breaststroke through the soil in search of grubs and slugs.

Although many people think of moles as rodents, they are insectivores, tiny little eating machines (averaging seven inches long) that consume nearly their body weight in insects, slugs, and grubs daily. Anyone who has ever plucked a slug or pinched a bug knows that they don't pack much weight, so imagine the number of victims it would take to equal the mole's $1\frac{1}{2}$ to four ounces.

Seven species of moles are found in North America, except in the arid and rocky soils of the Great Plains, Great Basin, and Rocky Mountains. A mole needs hospitable, moist soil, which provides it with both shelter and the steady diet of invertebrates it needs to fuel its racing metabolism.

These solitary, territorial animals are engineered perfectly for their underground lives. Their velvety fur, to which no soil can cling, lies flat against their bodies and doesn't hamper their forward or backward movement. Their eyes are tiny, hidden in thick pelage and sometimes covered with skin. They have no external ears, and their large, claw-tipped forefeet, with pink-tinged soles facing out-

ward (they look like they were mistakenly put on backward), are powerful earth movers.

As a mole breaststrokes through his dark, moist world, he pushes the soil out of his way and down the sides of his body, forming both feeding runways and tunnels. I think of the shallow feeding runways as ephemeral drive-in, or dig-in, eateries. The mole cruises along just below the soil's surface at the root level of many plants and uses his highly developed senses of smell, touch, and hearing, as well as sensitivity to vibrations, to locate his prey. As he burrows, he leaves behind a telltale wake of loose, crumbly earth. Moles sometimes only travel this mounded runway once before moving on to an area with a bigger payload of insects.

The permanent tunnels of moles are used year-round and are located about a foot underground. These tunnels often lead to the small (usually about six-inch) grass-lined burrows that by early spring may be crowded with three to six young. In the building of these deep, permanent tunnels and burrows, the industrious mole pushes the soil up to the surface and into its characteristic molehill.

The mole in my woodland border was doing double duty for me on the night I rediscovered him. He plied the earth for the pests that often wreak havoc in my garden, and as he delved he aerated

• If You Don't Want Them •

Yes, there are benefits—but there are drawbacks, too. Take care when walking or working around tunnels, so as not to injure yourself.

- To discourage moles, stamp down their tunnels and mounds with your feet. Put all plants back in place, then water the area thoroughly. Use a roller on the lawn.

- Don't over-water or over-fertilize your lawns, as this encourages grubs, slugs, and other mole favorites. Water deeply as needed. Instead of a diet of chemical fertilizers, sprinkle fresh grass clippings or compost onto your lawn.

- To protect small beds and borders, use hardware cloth. Bend the base of the mesh so that it forms an L shape. Dig a narrow trench 20-inches deep; insert the mesh with the L away from the bed, and backfill with soil, leaving 3- to 4-inches of the mesh above the ground.

and mixed the soil into the consistency of a perfect chocolate cake. As a side benefit, the cakelike earth had much better water absorption and a healthy mix of nutrients.

Today, as I wander through my gardens, I often imagine the subterranean yet parallel universe of the mole, who goes about his daily chores and life-and-death struggles just a few inches beneath my feet. The mole that visited my woodland borders did no damage to

- Put a barrier around treasured plants. Dig a trench 18- to 20-inches deep; fill with concrete, rubble, or stones (which moles cannot dig through); and cover with soil.
- Build a mole "mercy trap" to capture them alive. Dig to the floor of an active "run" and sink a coffee can into the soil so that the rim of the can does not protrude above the floor. Do not refill with soil; simply darken the run by placing a board across the hole above the ceiling. Remove the cover at least twice a day to check for moles. Relocate the captive to a new area.
- Stick inexpensive pinwheels into the soil near tunnels and mounds. Moles, with their delicate skulls, are extremely sensitive to vibrations.
- Bury a glass bottle upright in the soil near a run. Leave about an inch of the neck above ground. The weird whistling vibrations this produces disturbs the moles.

my gardens. It took only a minute for me to pat my tousled bulbs back into place and sprinkle them lightly with water. Both my garden and I were better for the mole's brief passage.

In Praise of
Garden Snakes

"Most of us need to be humbled more often, to be reminded that nature is not only more complex than we think, it's more complex than we can think."

—GARY PAUL NABHAN

This morning the sweet scents of summer flowers mingle with the unmistakable fragrance of the approaching autumn. Spires of brilliant goldenrod stud the grassy meadow that stretches from our porch to the edge of the bay. I wrap an afghan around my shoulders and head for the rock ledge where I like to sit and sip my first cup of coffee. Just a few feet ahead of me, I can see a narrow parting of the grasses, and slowly, a slender ribbon of garter snake (*Thamnophis sirtalis*) comes to rest on my favorite seaside perch.

Obviously, I am not the only critter seeking the warmth of the morning sun. I relinquish my ledge to the snake, slide into the old green Adirondack chair beside the bayberries, and sit back to watch. Not much happens at first, because the handsome striped garter snake (like all its kin) depends on the sun's heat to jump-start and maintain its bodily functions. Snakes are cold-blooded, or poikilothermic—their body temperature fluctuates with that of their environment. If a snake's body temperature is below 39°F, it can't move and is defenseless; if its temperature reaches 104°F, it perishes.

When I was a child, my handsome young Uncle Eric taught me that snakes were to be watched for and respected, but never feared. "There are snakes in these hills, Sharon. Always look before you step," he cautioned me as I ran outdoors and headed for the rocky trail that wound through the chaparral. Because of my uncle's quiet, matter-of-fact manner of teaching, I was always aware of the possibility of snakes, but was never afraid. Snakes were just another part of the natural world around me, as fascinating as trap-door spiders and centipedes and the reclusive horned toads that hid among the sandstone boulders.

The garter snake moved slightly and fixed its unwavering stare in my direction. I thought back to a summer years ago, when I sat patiently in my grandmother's sunny garden and practiced telepathy (to no avail) on the snake who lived in the wall near our apricot tree. My girlfriend had lent me a book about a tiny lady named Grace Wiley, who owned the "Zoo for Happiness" in Long Beach, California, just a few miles from our home. The sprightly Miss

Wiley was renowned for her handling of every type of snake, even the most venomous species. Her credo was that snakes will respond to respect, appreciation, affection, and good manners. People journeyed from all over the world to watch her in her "Gentling Room" as she charmed each reptile with a combination of silent communication, softly spoken praise, and a light stroking with a padded petting stick. As her unwilling and cantankerous subject relaxed, she moved closer and slowly caressed it with her bare hands. Incredulous visitors quickly became astonished admirers and believers, and I, though only acquainted with this remarkable lady through the printed word, became a lifelong fan.

Grace Wiley was probably one of the best ambassadors of good will ever to represent the beleaguered and maligned family of reptiles. No creature is more loathed or feared. In *Cultures of Habitat*, Gary Paul Nabhan states,

"Scientists at Wood's Hole reviewed experiments demonstrating that people from all corners of the planet have adverse physiological responses to sudden sightings of snakes..."

Nabhan writes that just the appearance of a snake produced

"greater changes in blood pressure and heart rate for a longer period of time than did images of knives, pistols, or automobile accidents..."

This deep-seated aversion is called ophidiophobia, and I have witnessed it firsthand. Last summer I presented a slide show and lecture called "Gardening for the Wild Things." As I projected photographs of beneficial garter, gopher, and king snakes onto the screen, three attendees bowed their heads and one promptly exited the auditorium. Later, while we discussed inviting wildlife into a garden setting, many in the group admitted that they would like to hide the welcome mat before these particular visitors arrived.

On an earth that was once ruled by reptiles (the dinosaurs), geologically speaking snakes are relative newcomers. In North Africa's Sahara Desert, scientists uncovered their fossil remains in the Lower Cretaceous rocks dating back 130 million years. Although I am a snake admirer, I must admit that a sixty-foot fossil, discovered in Middle Eocene rocks in Egypt, made even me squirm. Since a snake can swallow prey larger than its head, several times its diameter, and 50 percent or more than its body weight, such a gargantuan specimen would probably strike terror in the hearts of even our gigantic Maine moose.

A few years ago, I sat inside the snake exhibit at the San Diego Zoo and watched as a strikingly patterned king snake fed on a rodent. I'll skip all the details and get right to the point. Snakes have amazingly flexible mouths with bones on both sides of their jaws that can dislocate to handle sizable prey. While they are eating and digesting their meal, the sluggish, distended snakes are fair game for the animals who feed on them, but if they survive their feast they can fast for up to a year if necessary.

When I researched the subject of reptiles for one of my lectures, I learned that of the 2,400 species of snakes (more or less) worldwide, approximately 270 have venom that is harmful, but not necessarily fatal, to humans. David Dickey, herpetologist for the American Museum of Natural History in New York City, recently told me that of the twenty-three species found in the continental

To Attract Reptiles to Your Garden

1. Build a small mound of branches and brush in a corner of your garden.

2. Sink shallow saucers into the soil and keep them filled with water.

3. Place hollow logs in sheltered areas of your yard.

4. Malcolm Hunter, co-author of Maine Amphibians and Reptiles, built a rock pile and inserted a PVC pipe horizontally into the middle of the pile, leaving the ends of the pipe exposed. He then compacted soil over the rocks and pipe, taking care not to plug openings. This provides access to both reptiles and amphibians.

5. Make a "basking spot" by placing a flat stone in a sunny, exposed area near a hedge, rock wall, or border planting that will furnish shelter in case of attack.

6. Never use herbicides, pesticides, or rodenticides, which can be harmful to pets and wildlife, and will deter the things that snakes eat.

United States, most, but not all, secrete a toxin that is not especially poisonous. "Our venomous species are not among the top ten in the world," he said reassuringly.

I recently interviewed Malcolm Hunter, co-author of Maine Amphibians and Reptiles, and asked him for some words of wisdom about snakes. "Be nice," he urged. "In most parts of the United States, the odds of somebody having a venomous snake in a garden

· If You Don't Want Them ·

Note: Every month I receive a few letters about the "problem of snakes." So, for those of you who think I am a turncoat for writing this essay, here are a few suggestions.

1. Disregard my tips for attracting snakes.
2. Don't place birdfeeders close to your residence. Fallen seeds attract rodents, and rodents attract snakes.
3. If you live in an area that has species of venomous snakes, keep the grounds closest to patios or play areas clear of brush so any snakes can be easily spotted.
4. When all else fails: My uncle Willie Clarke, who ranched near Nogales, Arizona (snake country!), swore that his pet peacocks were the best snake patrollers in the world. Their shrill screeches alerted him whenever a reptile approached the ranch house, but beware: peacocks graze on tender green plants.

are about 100 to one." As for the other garden-dwelling snakes, they are just earning a living. Their diverse diets include many of the "pests" you work to eradicate from your garden. They'll eat everything from insects, slugs, leeches, and grubs to a plague of rodents and even other snakes. Some do consume eggs, birds, toads, lizards, worms, and frogs, but as I taught my son, Noah, "that is nature's way." Easy for me to say, but I am quite sure that if I saw a snake heading for one of my beloved toads or birds, I probably would be very unscientific and use peaceful diversionary tactics to deter it.

The old saying "What goes around comes around" applies to snakes, too, for they are often meals for other animals. Hawks dive from great heights to capture them in their talons. Foxes, martens, weasels, ferrets, fishers, wolverines, bobcats, badgers, and even some humans, who consider snakes a chickenlike delicacy, feast on them.

It is always a good idea to know which species of snakes inhabit your area. Learn to identify them with a local field guide (coloration and markings may vary greatly), and avoid handling them. Even non-venomous snakes are capable of biting. If it is necessary to move a snake in order to protect it, or for your peace of mind, consult Peterson's Field Guide to Reptiles and Amphibians. This informative guide devotes two pages to crafting and using a snake stick or capture noose.

My visiting garter snake must be warmed to perfection by now. It assumed a lazy S position and then slipped silently into the thick hedge of bayberries and roses that mantles the cliffs. I would love to get a closer look at this three-foot specimen, but my field guide informs me it is "quite pugnacious, tries to bite when captured, and may produce a strong anal scent when handled." I think I'll save all that pleasure and stimulation for another morning. For now, I am content to communicate a silent, Grace Wiley-style faretheewell to this garden dweller who is always welcome to share my favorite seaside seat.

A Blessing of Toads

*Never underestimate the power of a woman
or the usefulness of a toad.*

I t was a parade without fanfare—an occasion that could
have passed unnoticed—but for my family it was both a
blessing and a tradition we savored. On warm spring
evenings, we rushed through supper and walked quietly out
to the hedge by our garden gate to await the arrival of our
favorite visitors.

As night closed in around us, the line of street lamps bor-
dering the sidewalk flickered hesitantly, then flashed on. In

the round, white pool of light below our neighbor's lamp, we could see two dark lumps moving resolutely toward us. We watched silently, afraid to wiggle even a cramped toe, as the small forms ambled within inches of us and slipped under the garden gate.

The handsome Western toads (*Bufo boreas*) passed us and didn't acknowledge our presence even by a blink of their bright, golden eyes. On bandy legs, they shuffled down the pathway and into our gardens. We settled on the grassy bank above them, and though they were now out of sight, we could hear their chicklike peeping mingling with the cat-purrs of our resident spadefoot toads (*Scaphiopus hammondii*).

In the spring our pond was livelier than a singles' dance. The toads seemed to spend their nights grappling like wrestlers and calling to each other with a mélange of eerie grunts, chirps, moans, and trills. I know that for the toads it was music with a message, but to me it sounded like chaos.

When Noah was a child he told me, "The animals are talking to us; we just don't know how to understand them." I often think of this when I hear their strange cacophony. Toads have calls as distinctive to their species as do the birds, and many have repertoires to rival those of mockingbirds. Through their calls, they are able to communicate their specific species, territorial claims, availability, alarm, and aggression.

Although I don't understand every vocalization, some of them are so obvious that they leap all language barriers. One evening I watched as a plump, warty male climbed atop another and grasped it in a traditional mating clasp. Immediately, the toad on the bottom uttered a series of sharp, chirping, "release calls" that alerted the other male (and me) he was romancing the wrong sex.

D. H. Lawrence once wrote that he had never seen a "wild thing sorry for itself." Obviously, he didn't spend much time in the company of toads, who never suffer in silence. A toad in trouble, captured by a predator or caught in a trap, will scream and wail the most mournful sounds imaginable. These distress calls, which sound like a crying baby, convey a universally recognizable lament of terror and anguish.

By late spring, the grasses and algae in our pond were crisscrossed with gelatinous ribbons peppered with eggs. Within a few days, those black specks would hatch into tadpoles and begin the magical transformation into toads. The parents, landlubbers who seek water in which to mate and lay eggs, returned to their nocturnal routine of hunting and eating.

One summer night, armed with a flashlight, we followed a toad as he roamed under a bench and porch step in search of a meal. He paused and nuzzled aside leaves and twigs, and then, with a meteoric $^{16}/_{1000}$ths-of-a-second tongue flick, nabbed and swallowed a hapless beetle. The toad's long, sticky tongue is deadly—a gardener's dream and a bug's nightmare.

In a 1915 pamphlet entitled *Usefulness of the American Toad*, A. H. Kirkland published the results of his two-year observations of their

• How to Attract Toads to Your Garden •

- A moist, shady area that is a problem spot in your garden is a boon for toads. Plant therein an array of native, shade-loving plants, and mulch the area with a coverlet of leaves. Mulch provides shelter, moisture, and food for the toads.
- Sink a child's small pool, a half barrel, or a large bucket into the ground in your shade garden. Fill the bottom of the pool with rocks and soil, add water and a selection of floating and potted aquatic plants. For mosquito control, add some native *Gambusia* (mosquito fish), but don't add goldfish— they eat tadpoles.
- Don't discard damaged pots; recycle them for the toads. Chip out an entry hole in the rim. Sink the pot upside down 1 inch into soil with the entry hole facing south. Place the pot in a shady area that is not prone to flooding.

feeding habits. "Of the toads' total food," he wrote, "62 percent was made up of harmful insects. Should ants be included as injurious, as many housekeepers would think proper, this figure would be increased to 81 percent." Kirkland also noted that, "Toads fill their stomachs to capacity up to four times in a single night, accounting for as many as fifty-five army worms, thirty-seven tent caterpillars, sixty-seven gypsy moth caterpillars, or seventy-seven thousand-legged worms." Kirkland estimated that one toad can consume up

- Place some shallow terra-cotta plant saucers in your shade garden, and fill them with fresh water. Toads stretch out in shallow water and absorb moisture through their skin.
- Build a small mound of branches and twigs to shelter toads. The debris will also attract insects and slugs and provide the toads with unlimited snacks.
- Lay a terra-cotta pot or a hollow log on its side, and partially bury it in the soil. This "tunnel" is a toad resting spot.
- Dry rock walls are the perfect environment for toads, who find security, moisture, stable temperatures, and food in the crevices between stones.
- Avoid the use of slug and snail bait. Toads may inadvertently ingest the poison when they feast on contaminated victims.

to 10,000 insects during a spring and summer. Modern researchers have disputed Kirkland's liberal calculations and trimmed the number to approximately 2,000 insects per season, still a substantial dent in a backyard's insect population.

Celia Thaxter, author of the nineteenth-century classic Island Garden, wrote, "It seems to me the worst of all plagues is the slug." She used salt, lime, and wire cages to keep them from eating her beloved flowers, but the slugs were always victorious.

"Everything living has its enemy; the enemy of the slug is the toad," a friend told Celia and advised her to import some from the New Hampshire mainland. Celia promptly wrote a letter to a friend and asked for a supply of toads. Within a few days, a small box arrived and Celia released sixty toads into her Appledore Island garden. Throughout the summer she watched with satisfaction as her flowers thrived, and the toads grew fat on their plentiful diet of slugs.

Celia wasn't the only gardener who valued the appetites and abilities of toads. They were once in such demand that vendors sold them on the streets of London and Paris. An 1890s British newspaper article entitled "A Garden Friend" reported, "In such favor do toads stand with English market gardeners that they readily command a shilling apiece....The toad has indeed no superior as a destroyer of noxious insects, and as he possesses no bad habits and is entirely inoffensive himself, every owner of a garden should treat him with the utmost hospitality."

Last spring a friend rescued thirteen toadlets from a construction site and brought them to my gardens. What could be a better gift for an organic gardener than a Mason jar crowded with potential slug and bug busters? I knelt on the pathway, placed the container on the ground, and watched as the penny-sized toads climbed and tumbled their way out of the narrow opening. Within seconds they scattered, and some disappeared beneath the broad, velveteen leaves of a fragrant peppermint pelargonium. Three others shouldered their way into a thicket of forget-me-nots, while the rest scuttled into a bed of artichokes and 'Pink Panda' strawberries and vanished.

I pulled out my marking pen and scribbled a message on a large plant label and hung it in the middle of the entry arbor. Visitors who passed under the arch of 'Cécile Brünner' roses paused and read, "Walk gently please. We just released 13 tiny toadlets into the gardens, and we wish to treat them with the 'utmost hospitality.' "

As I prepared for a garden lecture a few months ago, I found my old copy of James Lipton's *An Exaltation of Larks*. This compendium of ancient collective nouns is fascinating. A "party" of jays, a "labor" of moles, a "charm" of finches, and a "parliament" of owls all seemed apropos, but a "knot" of toads disappointed me. The early semanticists who coined this term could have done better. Obviously, they never observed toads in parade formation, heard their songs of love and loss, or watched them stop a plague of locusts from destroying a garden.

I'm inclined to write Mr. Lipton a note suggesting that we alter this one term of venery to wording that expresses appreciation, respect, and fondness. I've experimented with dozens of phrases, but only one seems apt—a "blessing" of toads.

Skunk Scentsibilities

Get to know my favorite member of
Mother Nature's "cleanup crew."

I confess. Of all the scents associated with a garden, the pungent aroma of skunk is one of my favorites. Helen Keller once wrote that "Smell is a potent wizard that transports us across thousands of miles and all the years we have lived." Waft me even the slightest hint of *eau de mouffette*, and I am carried back through years and miles to the comfort and brilliant colors of my Grandmother Lovejoy's bountiful cottage garden.

Grandmother's yard was a favorite nighttime haunt of the paunchy striped skunk (*Mephitis mephitis*, which translates to "noxious vapor"), who claimed it as his own. Early in my childhood, I learned to respect this slow-moving member of the weasel family who made nightly forays through the flower beds. I never understood what he was doing or why he spent so much time in our garden, but I knew that wherever this cat-sized creature wandered, he left a lingering aroma that I could smell for days.

Skunks are placid, non-aggressive animals that waddle and putter their way through the world, protected by what naturalist Alan Devoe called "the enormous security of their skunkness." Most of them will barely notice a nearby person or animal and will continue on their way unconcernedly or pause to allow the intruder to turn tail. But if they are cornered by a foe, they are not afraid to stand their ground. When threatened, a skunk will arch its back, elevate and plump its bushy tail (under which the ammunition is stored), do a quick tap dance with its forefeet, and snarl or chatter its teeth. If these warnings aren't enough to discourage an aggressor (by this time I would be on the run), the skunk faces its enemy, quickly whips its rear end around until its body forms a "U" shape, aims, and shoots. A skunk can accurately hit a target up to twelve feet away, and if one shot doesn't do the job, it can spray another five or six times before emptying its musk glands. (The skunk's tricks work well for most predators, but the great horned owl is the exception. This large night hunter, with its four-foot wingspan, can silently swoop onto a skunk and dispatch it before it has a chance to spray.)

While being doused may not be the worst conclusion to an unpleasant encounter, the complex cocktail the skunk discharges does contain the active ingredient n-butyl mercaptan, a chemical so powerful it is now included in a number of popular animal repellents. The spray can temporarily blind a predator, cause nausea and nasal swelling, and will cling for months to feathers and fur (or clothing).

• Skunk Patrol •

Skunks as Yellow-Jacket Eradicators

If aggressive, ground-nesting yellow jackets (which are actually beneficial and prey on many insect and caterpillar pests) settle near your home or a play area and you want to eradicate them, a visit by a roaming skunk can solve the problem. Locate the entry hole to the nest (do this in the evening when the yellow jackets aren't active) and surround the hole with a thick circle of honey. Skunks will first be attracted to the honey, but once they discover the nest full of one of their favorite foods, they will tear it apart and eat the wasps and their brood.

The low-slung striped skunk, which looks as if it is wearing a black tuxedo with bursting backside seams, ranges throughout North America and is found in a variety of habitats including woods, grasslands, agricultural areas, cities, and suburban yards. Though they may live near our homes and in our yards, these retiring animals are not often seen. They sleep by day in tree stumps, woodpiles, under decks or outbuildings, and in burrows they have adopted or dug for themselves. They are nocturnal and solitary and accomplish their foraging, mating, and family-raising under cover of darkness. Sometimes the only thing betraying their presence in our surroundings is a rifled flower bed, an occasional tuft of hair, or the light, musky scent they use to attract a mate or mark their territory.

Following a brief winter dormancy, which is not a true hibernation, these adaptable, short-lived critters (average life span: two to three years) associate briefly during the mating season, from mid-February through March. After a gestation period of sixty to seventy-seven days, the female bears a litter of four to ten blind and deaf kittens with pink, crinkly bodies shadowed by their future coloration. At about six weeks, when the babies are fully weaned, the mother will lead her skunklets from the den and parade them single file into the world. For nearly a year, the youngsters will learn from her the ins and outs of "skunkness."

These pudgy omnivores, who are often referred to as Mother Nature's cleaning crew, are voracious feeders who eat not only household garbage, fallen fruits, and a legion of diverse garden pests, but also roadkill and any carrion they find. Depending on the bounty of the season, their nightly foraging may include crawfish, grasshoppers, potato bugs, hornworms, gypsy moths, grubs, armyworms, Japanese beetles, wasps, and an assortment of small vertebrates such as mice, voles, shrews, gophers (hurrah!), and young rats.

In two studies of the stomach contents of more than 3,300 skunks, scientists found that nearly half of the food consumed was insects (many of them considered injurious garden or crop pests), with a smaller percentage of mice, earthworms, grains, and fruits. They will also occasionally raid nests and feast on beneficials, such as turtles, snakes, toads, and lizards.

• Keep Skunks at Bay •

- Never leave pet food bowls outdoors overnight or hungry skunks will settle in for a meal.
- Skunks are garbage gourmets. Always keep garbage can lids tightly closed.
- Surround chicken coops with hardware cloth that is set at least 2 feet into the ground (skunks are great excavators), and bend the wire into an L at the bottom with the L facing outward.
- Block entry holes to basements, porches, and outbuildings with hardware cloth (using the same L-shaped bend), but be sure NOT to trap a skunk inside.
- To get rid of an unwanted skunk, the Bio-Integral Resource Center recommends that you look for entry holes and close all but one. Fill this remaining hole with loosely packed leaves or straw (when you see it's been disturbed you'll know the skunk has exited) and sprinkle flour or mason's chalk in front of the opening as a tracking powder. Return after dark to make sure the skunk has left for its nighttime peregrinations, then promptly seal the den with wire, metal or wood. Note: This should NEVER be done in March, April, or May when a litter of babies may be present, or in the summer months if you suspect that there may be a second, late litter in residence.

- Leave an ammonia-soaked rag or sponge in a bowl in any area where you want to discourage skunk visits.
- If you happened to be in the line of fire, Dr. Milo Richmond of Cornell University recommends that you wash sprayed clothing in strong detergent and rinse in diluted vinegar, orange juice, or tomato juice (don't use tomato juice on lightly colored clothes). If leather shoes are affected, bury them in clean kitty litter for a few days.
- If your pet is doused, don some gloves and immediately give the animal a sponge bath (never immerse it) of dilute bleach (1 tablespoon bleach per gallon of water) to remove odors, then wash with a pet shampoo. Some pet owners wash sprayed pets with straight tomato juice, dilute vinegar, or commercially available products that counteract the chemicals in the spray.
- The Humane Society of the United States recommends that you use a quart of 3 percent hydrogen peroxide mixed with $1/4$ cup of baking soda and a teaspoon of liquid soap to wash both your clothes and animals. Ellen Sandbeck, author of *Slug Bread and Beheaded Thistles*, had good luck washing her skunk-sprayed dog with soap and a vinegar rinse. She then rubbed baking soda into the dog's fur and rinsed again. She believes that the baking soda chaser was what made this concoction work.

• A Word of Caution •

Teach your children that there is no such thing as a "tame" wild skunk. If a skunk is behaving strangely, it may be ill or have rabies. Contact your local animal control agency immediately, and keep pets and children away from the skunk. Keep your pets' rabies vaccinations current.

By autumn, the skunks who ramble through my garden are plump and sleek from a summer of easy living and abundant food. Their stored fat will help sustain them through the lean months of winter. Occasionally, the skunks will emerge from their burrows or communal dens to fortify themselves with whatever they can scavenge.

Late these autumn nights, I can smell the familiar perfume, that calling-card of the skunk that takes me back to my childhood in a garden. When I go out to make my first rounds in the morning, I can see the pockmarked landscape of last night's visitor. Small excavations and disturbed mulch let me know that the skunk has been mining my borders for pests. Perhaps he will succeed where I failed in defeating some of the insects and grubs that attack my favorite perennials. The skunk may be part of the cleaning crew of the earth, but he is also my secret and tireless pest-buster, the head of the night shift that oversees the health of my garden. For this, and for that inimitable whiff of my childhood, I am grateful.

Confessions of a Worm Rescuer

You can't have enough of these
hardworking composters.

Worms wiggled their way into my life when I was a child growing up in the hills of Southern California. After each infrequent rainstorm, streams of mud, which glistened with hundreds of earthworms, coursed down the hillsides and onto our sidewalk and driveway. I became a dedicated worm rescuer, gathering them in a coffee can, then transporting them to the bountiful flower beds in my grandmother's garden.

Grandmother taught me that the worms worked magic as they literally ate their way through the earth. They left behind a vast network of slime-sided tunnels and rich mounds of castings (worm manure). Their excavations improved the soil's structure, increased aeration and root penetration, and allowed water to percolate slowly into the ground. Practical considerations aside, it was exciting to dig into the soil and find wriggling skeins of my rescued worms everywhere.

My early belly-to-the-ground observations never prepared me for the affection I would someday feel for my own "family" of worms—the bucket of red wigglers (and their descendants) which I received from George Kryder. Unlike the earthworms I had liberated into the soil in Grandmother's garden, these red invertebrates (Eisena fetida) had an appetite for garbage and thrived in the confines of a compost pile or worm bin.

As George shoveled knots of writhing worms from his bin into my bucket, he assured me that if I kept them protected, well fed, and happy (a happy worm?), they would multiply and turn our garbage and yard clippings into a rich "black gold compost" for our gardens.

I accepted the gift with a bit of trepidation. Did I really want the added responsibility of caring for thousands of worms? Weren't teenagers enough? As I strapped the worms securely into the passenger seat (my first stirring of maternal instincts for them?), I wondered how I would explain this new adventure to Jeff.

I rushed home and, with the help of my surprised husband and son, built a simple four-by-eight-foot stacked cinder-block bin topped with a removable lid of exterior plywood. The lid would protect the worms from predators (they are a favorite food of so many creatures) and keep their bedding area shaded and moist. Jeff poured a layer of shredded pine needles and leaves into the bin and I dumped in some edible kitchen garbage. Then we gently slipped the worms out of their bucket onto their new bed, added another layer of leaves and a sprinkling of water, and welcomed them to their new home. Life changed for the better after the worms arrived. Instead of being just an added responsibility, they became productive members of our family. Every morning I carried our garbage pail outdoors, raised the bin lid, and greeted the quiet multitudes as I fed them their breakfast. I was amazed at how quickly the worms (and bacteria) turned culinary disasters, unwanted leftovers, clippings, wet paper, and those fuzzy green things that lurked in the recesses of our refrigerator into food for our gardens.

I fill half of our worm bin at a time and alternate sides about every three weeks. Worms are like teenage boys: they follow the food. When they move to the newly-filled side of the bin in search of a fresh meal, I dip into the other half and shovel the worm compost into a large, covered trash can. Whenever I want to top-dress a border or a flock of my containers, I shovel mounds of the compost into my wheelbarrow and deliver it meals-on-wheels style to my fortunate plants.

When I first began to vermicompost (the unsavory appellation for letting worms eat your garbage), I was meticulous about removing every red wiggler from the compost before I applied it. I was sure the worms would starve if separated from their Mother Ship—the moist,

• Build a Cinder Block Worm Bin •

Cinder block worm bins are indestructible and freestanding, yet movable.

1. Pick a spot out of direct sunlight.
2. Set cinder blocks end-to-end with the open end up in a rectangular shape.
3. Add more rows until you reach your desired height, making sure that each block overlaps the two blocks below.
4. Place hardware cloth on the bottom to exclude burrowing critters.
5. Cover the bin with a piece of exterior grade plywood to keep animals out.

After you add your kitchen waste, be sure to cover debris with a layer of leaves, grass clippings, or wet newspaper.

dark bin they called home. One day I inadvertently dropped a small knot of worms into a terra-cotta pot planted with wildflowers. The worms immediately nosed their way under a layer of mulch and disappeared. That pot of escapees became a test area, which I watched closely throughout the summer and fall. The plants thrived and the soil became friable and dark, and every once in a while a few worms surfaced as if to assure me that they were fine. Now I wouldn't dream of container gardening without using my secret recipe of compost, a layer of shredded mulch—and a handful of worms.

Indoor Worm Composting

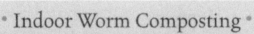

E ven apartment dwellers can compost with worms. Commercial worm kits are available at garden centers or through on-line and mail order suppliers or you can make your own. The worms will be content to live indoors if you keep them warm (55° to 77°F) and provide them with food and water.

1. Use any lidded plastic container, five gallons or larger.
2. Drill or poke ice pick size drainage holes into the bottom and air holes into the sides of the container.
3. Find a tray large enough to fit under the container to catch moisture.
4. To allow the container to drain freely, use any nonporous material to elevate it off the tray.
5. Add a layer of bedding (shredded wet newspaper or cardboard).
6. Add worms, topping them with another layer of bedding.
7. Feed worms household food garbage, including eggshells, coffee grounds (and filters), tea bags, potato peelings, and carrot tops. Always top with a layer of bedding after feeding. (Note: Foods not recommended for worm consumption are cheese, butter, meat, and oily products.)
8. Keep lid on container when not in use.

My friend Mary Appelhof told me that it takes at least eight gallons of water to rinse one pound of kitchen scraps through a garbage disposal. This is a waste not only of water, but also of precious garbage that could be recycled into unlimited possibilities for a garden. I cringe whenever I hear the sound of a disposal and am not too proud to beg for the kitchen scraps for my worm bin. This probably sounds like the proselytizing of a fanatic, but I know from experience that this urge to rescue unwanted and unappreciated garbage strikes everyone who vermicomposts.

Early this summer, our worms will travel with us on our cross-country trip to our cottage in Maine. I'll pack a pound of them (about 1,000 worms) into a large, insulated picnic cooler partly filled with shredded newspaper and soil, and belt them securely into the back seat. It always bothered me to waste uneaten food as we travel east, but this year I'll just lift the lid and drop our leftovers into the cooler. By the time we reach Maine, there should be a supply of castings to add to our container gardens and enough worms for a small bin. I think the worms will enjoy a summer in Maine. Now, I just have to figure out how to break the news to Jeff.

Holiday Feasts for Birds & Beasts

Nuts and berries,
Seed-filled flowers,
Feed the birds
Through winter hours.

—SHARON LOVEJOY

My tiny kitchen is a disaster. Mounds of dried flowers, gray-blue clusters of elderberry, tall tassels of coppery red sorghum, and plump discs of sunflowers occupy every square inch of our farm table. Miniature ears of brilliantly colored Indian corn and strawberry popcorn, and a variety of pinecones, acorns, and walnuts, are snuggled in a flat ash basket on the wooden counter. Cookie cutters, cooling racks, and a large yellowware bowl brimming with a "Songbird Supper" crowd my marble pastry slab. Anyone walking through the kitchen door right now would certainly find it difficult to believe that beauty and order will emerge from this chaos.

My Grandma Clarke, famous for her closets and drawers filled with presents, taught me to prepare for the holidays throughout the entire year. I cache gifts whenever I find something that's perfect for one of my friends or family members, and I do the same for my garden critters. I carefully observe the animals in my yard, note their special food preferences, then harvest and save those plants.

Last month I cut long, dry stems of cosmos, scabiosa, coneflowers, zinnias, coreopsis, and sunflowers, separated them into hand-sized bundles, and hung them from the rafters in my herb room. Though some people would consider these cuttings fit only for the compost pile, I knew from watching the animals that each dried flowerhead contained a supply of delectable seeds. Today I'll fashion these gleanings into wreaths, garlands, and thick sheaves.

Years ago I discovered a pile of century-old *St. Nicholas Magazines* in a bookstore. As I read through an 1880s edition, an illustration caught my attention. The drawing pictured a flock of birds feeding on a thick sheaf of wheat bound to a tall pole. The accompanying text explained this Norwegian tradition: "On Christmas Eve the sheaves are tied to the pole, and on the morn the birds come to feast."

Kary's Squirrel Cottage Cookies

Hang these nutritious treats from tree boughs for squirrels to nibble on.

Makes 2 dozen cookies
1 cup butter, softened
$1/3$ cup granulated sugar
1 egg
2 cups all-purpose flour
$1/2$ cup finely chopped nuts (toasted almonds, peanuts, walnuts, or pecans)
$1/2$ cup white cornmeal

Heat oven to 350°F. Cream butter, sugar, and egg together. Stir in flour and nuts and mix well. Shape into balls the size of small walnuts; place on ungreased cookie sheet. Place cornmeal in a small bowl. Dip the bottom of a drinking glass into the bowl, then flatten dough balls with bottom of glass. With a straw, make a hole near the center of each cookie. Bake for 14 to 16 minutes. Place cookies on a rack to cool. String twine or thread through holes and hang cookies from branches. (I sometimes cut a leafless tree branch, stick it into a crock filled with sand, and hang my cookies on the bare twigs.)

Recipe courtesy of Kary Blair Gonyer

Songbird Supper

3 cups bread crumbs (or leftover biscuits, donuts,
   or muffins)
1 cup wild bird seed
1 cup diced apples, berries, or raisins
1 teaspoon sand or commercial bird gravel
Suet or canola oil (just enough to moisten all
   ingredients until they hold together)

Mix together crumbs, half the seed, and all the fruit and
sand. Grind suet and melt in a double boiler. Cool suet,
then reheat until liquid. Add to dry ingredients and mix
thoroughly. Pour into a casserole or loaf pan. Pat firmly,
cover with foil, and refrigerate.

Celia Thaxter, one of my favorite nineteenth-century garden
authors, wrote,

> In the far-off land of Norway,
> where the winter lingers long and late,
> for the singing birds and flowers,
> the little children wait;
> All the stalks by the reapers forgotten,
> they glean to the very least,
> to save till the cold December,
> for the sparrows' Christmas feast.

• Log Bird Feeder •

1. Cut a log (about 4 inches in diameter) into a 12-inch-length.

2. Screw a heavy-duty hook into one end of the log.

3. In a random pattern, drill holes $^1/_4$- to $^1/_2$-inch wide and deep into the wood.

4. Fill holes with "Sharon's Super Energy Booster for Birds" (see recipe on page 132) and suspend feeder from a branch or eaves.

5. For gift giving, tie small bunches of acorns and berries and sprigs of evergreen to the top of the feeder with festive holiday ribbon. Include a bird book, and a jar of "Sharon's Super Energy Booster for Birds" and/or a recipe card with directions for making it.

I love the sentiment behind this tradition and do my own version each December 24th. Instead of a sheaf of wheat, I use the bundles of dried flower stalks from my herb room, bind them tightly with thick strands of sisal, and tie them onto a tall pole in my native-plant garden. Early each Christmas morning we awaken to the plaintive calls of the golden and white-crowned sparrows that are always the first to discover our gift.

Decorating a tree for the birds has been a family ritual for twenty-four years. In the beginning we simply garlanded small trees with strings of popcorn and cranberries. Now the trees are more elaborate and whimsical, and we not only decorate a tree for the birds, but also one for the squirrels. Jewel-like strands of golden raisins, berries, fruit slices, and holiday cookies in fanciful shapes drape the bird tree. Clusters of Indian corn, baskets of nuts and fruits, and my friend Kary's Squirrel Cottage Cookies trim the squirrel tree, though these crafty creatures always try to claim ownership of both.

After testing a simple homemade log feeder in our garden, we made some for our bird-loving friends this season. I stuffed chilled "energy booster" (see page 132) dough into holes drilled in a small log and suspended the log from an oak branch. Within a few hours the inquisitive chickadees sounded a chittering announcement and a cavalcade of birds appeared. Nuthatches, wrens, brown creepers, and a plain titmouse crept up and down the log, probing into the holes and picking out small chunks of the mixture. Just watching their antics made me happy. I felt content with the knowledge that the puttylike dough would provide my guests with a high-protein, energy-boosting food that would keep them warm.

My friend Kary will arrive soon with her traditional gift of cookies for the squirrels and birds. We'll decorate the trees, then tackle the mounds of dried goods on the table and counters. By

evening, the kitchen will be restored to normal and my old red wheelbarrow will be overflowing with decorations and gifts for my loved ones... including those with feathers and fur. Joyous and peaceful holidays to all!

Return of the Native

*Plied with catalogs of native plants, I can be
an armchair gardener on a cold winter day.*

I'm up to my earlobes in garden catalogs and feeling happy
and hopeful despite howling winds and lashing tree
branches. The weather breaks just long enough for me to run
outside and fill bird feeders, stuff suet treats in the chickadee
baskets, and clean birdbaths clogged with leaves and pine
needles. The pathways and beds are littered with cones and
twigs and broken branches. I head for the garden shed to grab

the rake and clippers, but before I walk ten steps the skies open and force me back indoors to the fragrant warmth of an almond-wood fire. Circumstances of nature excuse me from tidying the gardens and allow me the luxury of a long morning of reading plant lists, and the quiet companionship of my two old dogs.

I am always amazed by the quality of catalogs and the vast store of information each has to offer, sometimes for free, sometimes for just a few dollars. I am on a quest for native plants to integrate into my gardens at Seekhaven, my home, and for the native garden at my shop Heart's Ease, so my attention has been focused on information from the National Wildflower Research Center and Larner Seeds. I chuckle as I read the Larner catalog's introduction: "A plant is not a couch," writes Judith Larner Lowry, founder of Larner Seeds. "Many people are victims of plant anxiety. They want to know exactly what will do what and where. But gardening is not interior design, and the great part is, there are no absolute certainties…plants are not furniture." I realize that, an hour before, when confronted by my messy yard, I had been suffering an undiagnosed case of "plant anxiety," trying to keep my gardens as under control as my living room furniture.

Two plant lists are growing in the notebook perched on my catalog-laden lap. The list for Heart's Ease includes a rainbow tapestry of annual wildflowers and graceful grasses to dance around all of the native perennials planted last winter. The annuals would provide a rich source of nectar for butterflies, beneficial insects, and hummingbirds, and yield a cargo of seeds to feed hungry birds and to self-sow for future generations. My Seekhaven list includes additional shade-loving shrubs, as well as nectar and berry plants to charm the birds out of the sky and the critters out of the hills.

⁕ How You Can Plan for ⁕ a Return of the Natives

1. Open your eyes to native plants in your area. Observe their growing conditions.
2. Select your garden site and draw a basic plan.
3. Weed the site so natives don't have to compete with invasive exotic species.
4. Natives appropriate to your garden site may not need any soil amendments (the beauty of properly selected native species).
5. Establish plant communities by grouping species with the same site and soil requirements.
6. Diverse species of plants in a native garden will attract diverse species of butterflies, beneficial insects, birds, and mammals.
7. Native plants have already adapted to available water in their own habitats. But all plants, even those that are adapted to regions with prolonged periods of drought, must have their water needs met until they are thoroughly established.
8. Make use of the wealth of information in garden catalogs. Information is the fertilizer for exciting garden adventures.

A stack of catalogs and my lists slip to the floor as I reach for the binoculars on the windowsill and focus them on the old fountain, which is ringed with juncos and a sapphire blue Steller's jay taking turns bathing and preening. I scan the gardens and realize that my eleven-year investment of sweat equity, integrating natives into a fifty-year-old garden of exotic species, is finally paying dividends. Toyons (*Heteromeles arbutifolia*, a native evergreen holly), heavily laden and drooping under the weight of brilliant red berries, are alive with the comings and goings of plump, singing robins. Thickets of currants, gooseberries, coffee berries, and blackberries lure thrashers, jays, thrushes, and sociable coveys of quail to find shelter and food.

Elderberry bushes entice flocks of cedar waxwings from the sky to feast on their dusty blue berries. The waxwings (named for the spot of red on each wing) gorge on the berries and splash the ground with inky blue droppings before they move off in a noisy wave. The red-capped acorn woodpecker and smaller Nuttall's woodpecker probe their way up opposite sides of a native oak tree while a tiny, needle-billed brown creeper stitches an upside-down route in a crevice of Monterey pine bark. I settle back in my chair, pick up the National Wildflower Research Center list of recommended plants for my area, and add native scarlet columbines (*Aquilegia formosa*) for the hummingbirds, a Dutchman's pipe vine (*Aristolochia californica*) for butterfly larvae, and tidy tips (*Layia platyglossa*) for checkerspot butterflies.

I finish my plant lists and drawings, close my notebook, and restack the catalogs with a sigh of satisfaction. A stormy winter morning spent sowing dreams and ideas has yielded dozens of plant and habitat possibilities for future gardens: possibilities for conserving water and restoring the unique flora and fauna of my coastal foothills (starting in my own backyard); possibilities for creating habitats that will bid a congenial welcome to migratory birds and animals looking for their natural food supplies; possibilities for larval and nectar food sources for rapidly diminishing native butterflies, and for attracting native beneficial insects that will help to keep the gardens healthy without the intervention of herbicides, pesticides, and fungicides; possibilities for me to sit and enjoy the quiet complexities and interactions of a garden of natives without feeling the need to organize, arrange, rake, and groom. After all, "plants are not couches," and I refuse to suffer any more unproductive bouts of "garden anxiety."

A Pocket Full of Acorns

*Who can resist scooping up handfuls
of these autumn treasures?*

Acorns fell like rain on our rooftop last night. With each gust of wind, they pattered softly, like fat spring drops, then rolled down the steep incline and onto the pine needle-cushioned ground. I lay in bed half asleep and thought of the Native American Concow thanksgiving song, *Hu'-tim yo'-kim koi-o-di'*, "The acorns come down from heaven."

Autumn is my favorite time of year. The red oaks (*Quercus rubra*, the Celtic *quer* for "fine" and *cuez* for "tree") surrounding our cottage are turning deep shades of russet, gold, and ocher. Their acorns, the magical fairy toys of my childhood, are strewn across the mossy ground and along our roadway. I can't take ten steps without stopping to gather them by the handful.

When I first discovered acorns (from the Indo-European root that means fruit or berry), I thought that someone left behind a collection of wooden toys. I remember how I turned them in my hand and felt their smooth, hard shell and the tight-fitting cap perched on top. If someone had told me that the acorns I stuffed into my pockets were the product of the huge oak trees skirting the Arroyo Seco, I never would have believed them.

Years passed before I finally, aided by a patient Brownie troop leader, connected the acorn to the oak, and, just as important, the oak to the tiny acorn. I felt as though I had unearthed a buried treasure when I learned that nestled inside an acorn are the embryonic beginnings of what naturalist John Muir called "a broad and bossy oak tree."

Botanists estimate that there may be up to 500 species of oaks worldwide, with perhaps sixty in the United States. Oak trees are flowering plants of the Fagaceae family, whose female flowers, when successfully wind-pollinated by male flowers, produce acorns—with each species sporting its own distinctive shape and cap. Although oaks may live for as long as twenty years before they begin producing, they can remain prolific for centuries.

• Gather Ye Acorns While Ye May •

Each fall I collect baskets of acorns, some for planting, others for toys or feeding our birds and animal visitors.

For the Birds

Gather fresh acorns and store those you wish to save for winter bird food in a clean, covered container. Scatter them onto platform bird feeders throughout the season.

Grow a Mighty Oak

Plant acorns in deep containers (a 3-inch seedling may have a 20-inch root) with good drainage. Set each acorn on its side in a 1-inch-deep hole. Cover with soil, mulch lightly, set outside, and water sparingly. When the first set of true leaves appears, brush the mulch away from the stem (to prevent fungus and rot). Water as needed.

When the seedling reaches 2-inches tall, plant it in a sunny, well-drained area, remembering to take into account the eventual size of the tree. Keep mulch, weeds, and grass away from the seedling and protect from mowers and weed whackers. To protect young oaks from browsing mammals, form a tepee or cloche of chicken wire around the tree.

The cotyledons, or seed leaves, tucked inside the protective shells of acorns store carbohydrates, proteins, and fats that nurture newly germinated seedlings until the first true leaves appear and begin the process of photosynthesis. These same elements that sustain a young oak seedling made the acorn an invaluable foodstuff for many Native American tribes, who relied on this high-energy seasonal harvest as a staple. Acorns, with their tough outer husks, are perfectly packaged for storage, and once dried and kept from the elements can remain palatable and nutritious for up to 30 years.

During *Huwol-chukmol* (the Pomo name for the month the acorns fall), the ground beneath oak trees abounds with life. Birds—from the large, ungainly turkey and grouse to the tiny $5^{1}/_{2}$-inch Carolina wren—feast upon the bountiful windfall crop. Mammals, who must build up reserves of fat before the onset of winter, seek out the rich carpet of nuts (which may constitute up to half their diet) and feed heavily. Many of them will return throughout the winter and search through drifts of snow and leaves for any remainders. Rodents and a few species of woodpeckers collect sizable caches of acorns to carry them through the seasons.

Jays, intelligent members of the family Corvidae, gather many more acorns than they can consume, then stash (and often forget) them throughout the garden. Their forgetfulness is credited with new generations of oak trees that would not survive the competition for nutrients and sun if they remained near their parent tree.

This year the National Arbor Day Foundation began a four-month polling process to choose America's national tree. I was torn about how to vote. I love sycamores, which didn't even place in the top twenty-one choices, and I am awed by redwoods. But I felt that

• A Solid Investment •

According to the National Arbor Day Foundation, over a fifty-year period a single healthy, mature oak tree can generate $62,000 worth of air-pollution control, recycle $37,500 in water, and prevent $31,500 in soil erosion.

the oak, which is such a life-giving part of our environment and grows in so many states, should probably get my vote.

I did my homework and found that oak has been used to make paneling, furniture, tools and tool handles, wagon parts, railroad ties, houses, eating utensils, tableware, medicines, dyes, charcoal, cabinets, axles, Roman aqueducts, bridges, and boats (reputedly Old Ironsides' illustrious strength came from oak).

The Norse built the hearts of oak into the hulls of their boats. They believed that with this magical protection their boats would survive every misfortune. Druids, Greeks, Romans, Teutons, and Celts held the oak to be sacred, and the traditional Yule log was always of oak. Ancient cultures believed that sleeping under the shade of an oak tree would cure many ills, and that "...men fed with oaken mast [acorns], the aged trees themselves in years surpassed."

The oak won the title of National Tree with 101,146 votes out of a total of 444,628, leading the race from the first day of tallying. It wasn't its usefulness or deep-rooted history that made me cast my vote for the oak. The acorns persuaded me—the irresistible exuberance of the acorns that I stuff into my pockets each glorious autumn.

Fragrant Memories

"*Sweet perfumes work immediately upon the spirits for their refreshing; sweet and healthfull ayres are special preservatives to health and therefore much to be prised.*"

— RALPH AUSTINE, *The Treatise of Fruit Trees*, 1653

The golden sunlight of autumn glimmers off the indigo waters of John's Bay and flickers through the woods sheltering our Maine seaside cottage. I can still comfortably sip my daybreak cup of coffee on the old porch facing east toward Pemaquid Point, but a cutting breeze and the brilliant, paint-box colors of the birches, maples, and oaks signal the turning of the seasons.

Today we'll begin the process of packing and readying our home for a long, harsh winter. Window boxes and terra-cotta pots, crowded with fragrant and everlasting plants for my potpourri blends, are still defiantly brimming with blooms. We must harvest the final flowers and herbs, empty the pots and boxes, and store them inside until next summer.

Cartons and suitcases lie open on the living room floor. The assemblage of goods making the journey back to California is an odd one: the line of bleached seashells from the windowsill; bags and baskets of mosses, lichens, acorns, pods, leaves, and cones; sweet, colorful flower petals; and a mound of pungent fir and pine boughs.

Some travelers buy postcards and souvenirs, and take photographs of their favorite places. I like to carry home the colors, textures, and scents of my journeys and blend them into sachets, potpourris, and gifts that conjure up joyous memories of walks along Sea Lavender Cove, the mingled scents of balsam fir and roses, and the windswept bluffs of Monhegan Island.

Although I can't remember when I first began to collect and blend my own botanicals, I do vividly recall trailing along behind my grandmother as she harvested her garden's bounty, hung bunches of herbs to dry in the pantry, and spread roses and lavender across her screened cooling shelves. Sometimes, when I am stirring together petals, spices, and fixatives, I catch the slightest whiff of a familiar scent and am immediately back in my grandmother's garden.

Our sense of smell has the power to stir our emotions, transport us backward through time, and alter our taste and energy levels. Recently, researchers at Yale University's Psychophysiology Center

• Make Your Own Potpourri •

The joy of blending your own potpourris and sachets is that you can do it in a garage, workshop, or better yet, at your own kitchen table. You will need lidded jars for storage. Purchase a variety of containers, or you can simply wash out that old mayonnaise jar and fill it with your ingredients. Large mixing bowls are a must for blending. Go to a flea market or tag sale and buy some inexpensive but charming crockery or glass bowls (avoid using wooden bowls as they absorb oils).

Collect baskets, bowls, crocks, and boxes for your potpourris. Assemble a collection of old spoons to use in the blending and measuring of ingredients. Eye droppers simplify oil blending. Pharmacies and herb stores usually keep these in stock.

For processing herbs, seeds, citrus peels, spices, resins, and fixatives, a food processor is helpful, but not necessary. You may use choppers, peelers, corers, zesters, a nutcracker, knives, coffee and spice grinders, or an old-fashioned mortar and pestle.

and the University of Cincinnati offered scientific evidence that certain fragrances can reduce stress, increase alertness and efficiency, lower blood pressure, and affect the learning abilities of children and of patients suffering from Alzheimer's disease.

Potpourri Ingredients

Every potpourri should be an artful blend of flowers, leaves, spices, seeds, woods, and fixatives. Fixatives absorb the scents of the botanicals and oils, and preserve them for an extended period. Fixatives can be as traditional and exotic as tonka beans, orris, or calamus, or as common as fiber cellulose, which is dried, chopped corncobs. These elements of potpourri not only provide a mosaic of color and texture, but also are essential to the lasting fragrance of your blend.

Time is of the essence in creating potpourris, for the oils, fixatives, spices, fillers, and botanicals need time to blend and mellow. Don't rush this process or you will create "rotten pot," which is the literal translation of potpourri.

Gardeners have known for centuries about the curative and restorative powers of scent. The sixteenth-century essayist Montaigne said it most beautifully, "Physicians might, in my opinion, draw more use and good from odours than they do. For myself, I have often perceived that according unto their strength and quality, they change and alter and move my spirits and make strange effects in me."

Nowadays, potpourri and perfumed goods are ubiquitous. I've even found these products on the checkout counter at a car wash. But tracing their history back through the ages, I've discovered that

• Gather •

Start eyeing windfalls with a new appreciation. Collect acorns, leaves, lichens, moss, seeds and seed heads, pods, cones, shells, pieces of bark, rose hips, dried Indian corn kernels, and twigs. These ingredients will give your potpourri texture and visual appeal, and will reflect your personality.

• What to Grow •

- **Annuals** for color.
- **Bulbs** for color (most bulb flowers must be dried in silica gel to retain color).
- **Everlastings** for color, shape, and texture.
- **Herbs** for fragrance and varied texture.
- **Perennials** for color and shape.
- **Roses** for fragrance.

our myriad uses of perfumes, potpourris, and ointments can't hold a scented candle to the many ways our ancestors added fragrance to their bodies and homes.

More than 3,000 years ago, the Babylonians and Assyrians burned aromatic plants to drive away demons, purify the air, and restore health. Cretan athletes anointed themselves with oils before their competitive games, and gladiators wore perfumes. Persians built their homes of fragrant, incensed woods, and mosques were constructed with rosewater and musk mixed into the mortar. In the royal courts of ancient Japan, incense was kept burning twenty-four hours a day, but the scent was changed every fifteen minutes.

The Egyptians were the first to use fragrance in burial, embalming, and in their religious cults. They invented the art of the bath, soaked themselves in aromatic oils, and used perfumed body and hair oils. One of my old herbals describes an Egyptian socialite, who on special occasions wore a cone of wax and scented oil on the top of her head; as the cone slowly melted, it covered her hair and shoulders with a sticky, albeit fragrant, coating of goo.

Early writings of the Greeks and Romans describe banquet halls carpeted with roses, rose water bubbling in fountains, awnings steeped in rose perfume, mattresses and pillows stuffed with rose leaves and petals, and elaborate, richly perfumed boxes of botanicals placed throughout their homes. Greek herbalists extolled the virtues of fragrance, and recommended scenting bodies with "mint for the arms, thyme for the knees, cinnamon, rose, or palm oil for the jaws and chest, almond oil for the hands and feet, and marjoram for the hair and eyebrows." Can Elizabeth Taylor top that?

Monastic records of the Middle Ages indicate that sweet-smelling

• Garner •

Spices and seeds give a potpourri pizzazz. They add their own signature scents and provide an eye-pleasing array of shapes and textures. Reach for the cloves, cinnamon, nutmeg, star anise, coriander, and allspice. Experiment and use sparingly as an accent.

herbs and flowers were planted near and hung in infirmaries to promote quick recovery. Perhaps our present-day custom of sending flowers to the ill springs from this early practice.

During the reign of Elizabeth I of England, there was extravagant use of herbs and perfumes. The queen wore gloves and shoes scented with ambergris and cloaks redolent of gardens. She required her ladies-in-waiting and her pets to be heavily perfumed so that she would always be "enveloped in eddies of fragrance."

Elizabeth's subjects followed in her stead and used herbs and sweet scents to mask their body odors (no bath a day during the sixteenth century), and for perfuming linens, clothing, snuff, cosmetics, medicines, syrups, and vinegars, potpourris, and "sweet bags." Not content to have only one fragrance of potpourri in their homes, they took great care to change aromas several times a day.

In France, Louis XIV (the "Sun King") kept his rooms filled with potpourris, perfumed his body and clothes, and invented a new scent every day. Louis XV drenched doves in scented oils and released them at his dinner parties. In England, James II hired a "royal strewer" whose job was to scatter lavender, thyme, rushes, and marjoram over the castle floors. Sometimes, the strewer was so busy she needed as many as six helpers.

I could use those helpers right now. Packing such a strange array of keepsakes is a big job. But once I arrive home, I know I'll be thankful to have my workroom stocked with all these ingredients. I look forward to the slow, peaceful process of infusing oils, blending potpourris, and stitching together handkerchiefs and scraps of antique fabric for Elizabethan sweet bags.

I haven't broken the news to Jeff that we now have enough rose petals and balsam fir to follow the Greek tradition of sleeping on a mattress of "sweet repose." I'll just slip him an antihistamine and quote a fifteenth-century herbalist who believed that "sleeping upon the sweete smelling herbes of the fields and gardens will instille vertue, lulle, comforte and increase our dreams."

Conversations with Sunflowers

You'll never be lonely with these cheery,
inviting plants in your life.

B y springtime my old harvest basket is filled—not with
flowers, but with the sweet promise of those to come. I
store an ever-burgeoning supply of seed packets in the basket
so I can thumb through them a few times a day while I plan
and dream about my bountiful summer gardens.

Half of my seeds will be planted in containers on the terrace in my wild California garden; the other half will be
shipped to my friend Liz Kellett at Wind Song Herbs in Wal-
pole, Maine. Liz starts many of the seeds for my seaside porch

garden in Maine each year and faithfully tends the young plants until I arrive. She always seems relieved to see me, and helps load the thriving pots and flats of flowers into my car like a tired baby-sitter returning a passel of still healthy and undamaged children to their mother.

This year my shipment to Liz was mostly seeds of sunflowers in glowing shades of orange, gold, yellow, and autumnal hues of bronze, brown, mahogany, red, and rust. I chose these plants not just for their colors, but also for their range of heights and their adaptability to the sometimes cramped conditions of life in a terra-cotta pot. But even these sterling qualities would not interest me if these whimsical floral personalities weren't great magnets for the creatures I want to entice onto my tiny Maine porch.

Through every stage of their existence, sunflowers appeal to wildlife. Their succulent green foliage and protein-rich seeds are favorite foods for sixty species of birds, unimaginable numbers of insects, and thir-teen mammals—fourteen, if I include my shep-herd dog, Una, who loved to nibble the plump, green buds just before they burst into bloom.

Three summers ago, when I first shared the tight quarters of our Maine porch with dozens of sunflowers, I became aware of the endless stream of traffic they generate both day and night. During the early mornings bold little red squirrels ran beneath our occupied porch rockers and climbed the stalks to nip bites from

• Growing Tips •

Begin sowing sunflower seeds during April—a month the Native American Hidatsa tribe referred to as *Mapi'-o'cë-mi'di*, or "sunflower-planting-moon"—in small paper pots that break down rapidly when planted in the ground. The Hidatsas planted as soon as they could work the soil. The sunflower was the first crop planted, and the last to be harvested.

Protect young seedlings from cutworms with a "collar" fashioned from a paper cup. Remove the bottom of the cup, and then nestle it over the sprout, pushing it into the top 1-inch of soil. You may wish to use copper strips as collars; these will also deter slugs and snails.

Birds, squirrels, raccoons, woodchucks, chipmunks, and mice enjoy a meal of tender young seedlings. Cover the small plants with the lattice-work baskets often used for berries and small vegetables. Anchor the basket to the ground with a bent piece of wire or a twig.

Fertilize sunflowers with a half-strength solution of liquid kelp and fish emulsion every time you water; dress the soil with well-aged manure.

Refrain from using any pesticides, which can injure or kill many of the creatures attracted to sunflowers.

• Flights of Fancy •

Living fences: If you have plenty of space in a sunny location, you may want to plant a living fence of 'Russian Mammoth', 'Giant Gray Stripe', 'Paul Bunyan', or 14-foot 'Kong'. For a longer-term hedge, a dependable performer is the perennial Maximilian sunflower (*Helianthus maximilianii*), which will return each year and produce dozens of yellow flowers on 10-foot stems. The Native American Hidatsa tribe of North Dakota edged their fields with sunflowers planted in small hills spaced a few feet apart. They mounded the soil, tucked three seeds into the mound "at the second joint of a woman's finger," then patted the soil firmly into place.

Buckets of sunshine: Plant the diminutive 'Teddy Bear', 'Elf', 'Sunspot', and 'Music Box' mix—which range in size

developing flower heads. Goldfinches flew in throughout the day and pecked small holes in the leaves, as though they were partaking of a giant green salad. These same birds, joined by chickadees, jays, nuthatches, and white-winged crossbills, visited again in the fall to feast on the bounty of sunflower seeds.

When the sunflowers, with their nectar-rich central disk, began to flaunt their winsome faces, the cast of characters that performed on our porch changed dramatically. Monarchs, eastern checkerspots, orange crescents, swallowtails, painted ladys (who deposited single, pale green eggs on many of the leaves), and little from 2- to 4-feet tall—in large buckets, commodious window boxes, or terra-cotta pots. Provide them with rich soil (bagged potting mix is great), and top with a 2-inch layer of compost or mulch (but keep it away from stems) to retain moisture.

Sunflower dollhouse: Make a wooden box 18 inches wide by 24 inches long and 4- to 6-inches deep, with drainage holes. Fill with potting soil. Along the edge, plant the seeds of the tiny bronze-and-gold 'Sundance Kid', which grows only to a height of about 18 inches; leave an opening for an entryway. Water faithfully, and when the sunflowers are at least 6- to 8-inches tall, you can begin to furnish your dollhouse with a moss carpet, furniture of wood rounds, cradles made from walnut shells and milkweed pods, and acorn-cap stools.

wood satyr butterflies drifted lazily between the porch railings, then settled atop the broad faces and probed the disk flowers for their cargo of nectar.

The pair of ruby-throated hummingbirds who built their tiny nest in the oak near our kitchen door felt they should be the sole beneficiaries of the sunflowers' offerings. They zoomed around the porch and hovered over the butterflies until they dislodged them, then crisscrossed the flower faces and dipped repeatedly into their blooms. Only the gold-belted bumblebees were able to withstand the hummers' scolding attacks. They sat like furry brown bears at a pot of honey, and sipped nonchalantly while the hummingbirds did their best to annoy them.

Pollen-eating beneficial insects and the pale crab spiders that feast on them joined the ranks of those who shared the sunflowers during the long Maine days. By night, under skies faintly lit by the brilliant Perseid meteor showers, we discovered scores of subtly colored moths crowded onto every inch of the sunflower faces. They nectared silently, undisturbed by the beam of our flashlight, and only took wing when a bat on a hunting foray swept into their midst.

As I greet my cheerful sunflowers each morning, I often think of the seventeenth-century John Rea poem: "Into your garden you can walk and with each plant and flower talk." Like a mother with a newborn baby, I coddle and groom them, and carry on long, one-sided conversations. I suppose these crowded pots of flowers are my "children," and such good ones. They behave well despite sometimes less-than-perfect conditions, they seldom catch a bad bug, and they entertain me with their easygoing personalities and their constant parade of visitors. All this, and so much more, from a basket full of seeds and a kind friend's loving care.

The Ladies on My Windowsill

With scented pelargoniums, the slightest touch releases the fragrance of roses and spices.

"Good morning, ladies," I greet the bevy of plants that parade along my windowsill like a dance troupe of green-skirted ballerinas. The light shines through their parasols and petticoats of leaves in a myriad of subtle greens, grays, and golds. As I brush past them the mingled scents of nutmeg, apple, and lemon fill my sunlit studio.

Two Experts' Favorites

V. J. Billings of Mountain Valley Growers and Judy Lewis of Lewis Mountain Herbs, with a combined growing experience of forty-plus years, recommend these scenteds for both indoor and outdoor growing. Caution! Scenteds, which are South African natives, thrive outdoors only in Zones 9 to 11; in other zones, treat them as annuals. Take cuttings in the fall and share the winter with them indoors; plant outside after all chance of frost is past.

V. J.'s choices

'Attar of Roses', 'Rober's Lemon Rose', and 'Peacock Rose' are a robust trio of scenteds that will respond quickly to indoor TLC and produce small pink blooms, as delicate as the tiniest butterflies. Their leaves have great flavor for cooking, and also look terrific as garnishes. Outdoors these plants can reach 2 to 3 feet, ideal for a low hedge.

Nutmeg-scented (*Pelargonium x fragrans*) is perfect for containers and blooms prolifically throughout the seasons. Outdoors, it looks especially attractive when planted in masses or flanking pathways. V. J. notes that a large grouping of these white-flowering plants looks luminous in the moonlight.

Crispum lemon (P. *crispum* 'Lemon'), with its stiff, crinkled leaves and lilac-rose flowers, adapts well to topiaries and thrives in pots (it often gets lost in outdoor plantings). This scented can reach a height of 2 to 3 feet. The leaves are pungently lemony and are great in bouquets and cooking.

'Village Hill Oak' blooms constantly and has a distinctive chocolate blotch in the center of each deeply cut leaf. Plant this in a terraced garden and watch its blooms cascade over the walls.

Judy's picks

Gray Lady Plymouth (*Pelargonium x asperum* 'Lady Plymouth') has a warm, rosy scent and sports elegant leaves edged with a creamy variegation. These look handsome in borders, are great cut for bouquets, and can be used in potpourris.

Apple (P. *odoratissimum*), with its delicious green-apple aroma, does well in hanging baskets. Given the right lighting, this plant will be laced with tiny, snowflake white blooms.

Clorinda (P. *x domesticum*), with its upright, showy growth, bright pink flowers, large leaves, and potent eucalyptus scent, deserves the appellation "lusty." Although you can start this plant indoors, move it out to the garden for its best showing, then move over—it grows exuberantly.

'Mrs. Kingsley' (P. 'Mrs. Kingsley') is a parsley look-alike, and Judy uses its leaves in lieu of parsley as a special garnish. This plant stands upright and flaunts brilliant cerise flowers.

Peppermint (P. *tomentosum*), with scented, downy leaves, holds a special charm for Judy and me both. My son gave me a tiny plant for Mother's Day years ago, and it now fills a four-foot area of terrace and wall with its velvety green leaves and white flowers that look like a miniature formation of flying snowy egrets.

These old companions, a cherished collection of scented pelargoniums (Greek for "stork," which describes their stork-billed seed capsules), sometimes incorrectly called geraniums, have been my faithful roommates and garden dwellers since I had my first apartment in the attic of an 1880s boardinghouse. The "scenteds," once the most beloved and common of Victorian houseplants, must have felt right at home, crowded onto a shelf under the broad windows that overlooked an old-fashioned garden and a busy street.

"Beware," I was once cautioned by a nurserywoman with a vast collection of scenteds. "You can get hooked, and there's no turning back." She was right. First to set up housekeeping with me was a large 'Rober's Lemon Rose' (*Pelargonium graveolens*), which wanted to inhabit every inch of my sunny window. Then came the nutmeg (P.

x *fragrans*), with tiny, glaucous leaves and miniature white flowers, followed by a pendulous, ruffly apple (P. *odoratissimum*), content to spill like a waterfall over the edges of my narrow plant shelf.

The scenteds and I grew together. When I finally had a cottage with a sheltered patio, they spent their summers outdoors, luxuriating in the dappled sunlight and the sea fog of Santa Barbara mornings. They thrived despite their sometimes cramped conditions in terra-cotta pots, and when they finally were potted up as far as I could afford (both space- and money-wise), I clipped their tips for propagation and slipped them out of their containers and into the ground—where they flourished here in Zone 9 (they are hardy in Zones 9 to 11).

Years later, when I planned a community garden in the center of our village on the California Central Coast, I included a pathway of scented pelargoniums to be planted in front of a silvery cedar fence. In preparation, I added yards of dark topsoil and a blanket of humus to the border to give the scenteds the friable, quick-draining soil they require. By the time I finished working, I could slide my fingers into the earth as easily as I once dipped my hands into the chocolate cake my Grandmother Clarke left to cool on the kitchen counter.

When the scenteds arrived, I was a little depressed. A budget as tight as a corset necessitated buying small, actually minuscule, plants. All sixty of them fit into one cardboard box sectioned off like an egg carton and filled with tall, finger-thin pots. Within hours, we dug slender holes, set the plants firmly into place, and watered them. As I poked the wooden

• Caring for Scenteds Indoors •

- Plant scenteds in a light, fast-draining, sterile potting mixture.
- Provide the plants with at least 4 to 6 hours of light daily (either sunlight or artificial).
- Don't leave them too close to a window if temperatures become extremely cold.
- From late March to October, fertilize them monthly with a natural fertilizer such as fish emulsion and kelp or fermented salmon.
- Turn the plants weekly for uniform growth.
- Pinch back growing tips to encourage fullness.
- Remove any brown or diseased leaves and dispose of them.
- If whiteflies or spider mites attack, simply put the

identification labels near each specimen, I shook my head: the signs were ten times bigger than the plants.

One of the finest attributes of the gardener is an unbridled sense of optimism. Who but an optimist would scatter seeds and expect flowers, or, even more unlikely, a giant pumpkin? Who but an optimist would stick a slip of plant the size of a pencil into the soil and expect an exuberant hedge of scented pelargoniums? The optimist in me was rewarded. Within a few months the long hedge was an undulating green mound, like a dragon dozing contentedly in the lee of the fence. That bountiful pathway gave me such joy, plants in your sink and spray them thoroughly with warm water.

- Allow the plants to dry out to a depth of at least an inch before watering, and never allow them to sit in a tray of water (elevate them on pebbles).
- If you over water and are plagued by fungus, aerate the soil with a hand or kitchen fork and water the plants with a strong (ten bags per teapot) solution of room-temperature chamomile tea.
- **Propagation:** I root most of my scenteds from cuttings, either in small vials of water or pots filled with a soilless potting mix. Judy Lewis uses the ready-to-go peat pots to which you add water. Spring cuttings root within two weeks; the fall cuttings may take up to four weeks.

and I, in turn, was able to share that joy with visitors of every age. Whether I was snipping plump handfuls of fragrant cuttings to share with an aspiring gardener or telling the stories of the plant's myriad uses in cooking and perfumery, the scenteds worked their way from my hands and heart to those of other like-minded souls.

The "ladies," with the sun shining through their green parasols and petticoats of leaves, continue their still ballet on my windowsill. I spend just a few serene minutes every day grooming them, pinching back tips to encourage fullness, watering only when the soil is dry, and giving them a pirouette turn each week to encourage even growth. Every perfect leaf becomes a mini harvest—some of them are tucked into my old sugar bowl to add flavor, some are baked with cakes, and others are dried on screens and added to my blends of fragrant potpourri. Even now, in the chill of winter, when the days are all too short and the long nights lie like a star-flecked coverlet over the earth, their sweet scents provide a gentle promise of spring.

Reveling in Rugosas

Think you can't grow roses?
Try these easy-to-please beauties.

Rugosa roses come from an ancient, beleaguered, and finicky family of plants, loved by many and shunned by the cowardly or lazy. Until recently, I was one of the cowards, but I've been converted and spoiled by the undemanding, silky-petaled Rosa rugosa, which thrives in the most improbable circumstances. The rugosas that first won me over grow among the pleated ledges of seaside stone in front of my Maine island cottage. They face the wild Atlantic Ocean defiantly, and when everything else is suffering from the heat and humidity of the dog days of summer, the rugosas are in their glory. They flaunt their fragrant pink petticoats, plump the succulent red hips (fruits) that earned them the nickname sea tomato, and put a spit-and-polish shine on their crinkled green leaves.

I think the small thicket of rugosas near my cottage is one of the reasons I fell in love with this place. On warm days their rich perfume wafts into my workroom and tempts me outside. I visit the roses and watch as the fuzzy black bumblebees roll around in the flowers and emerge with a dusting of golden pollen on their bodies. Sometimes early in the morning, the wary red fox sprints across the yard, stands up human-style on his slender rear legs, and plucks the fat hips as deftly as a gardener would pick an apple.

During nesting season, the prickly rose canes protect an array of birds from predators, and each night through all the seasons dozens of birds, betrayed only by their farewell-to-day peeps, fly in to roost. It always amazes me to see the sparrows, thrushes, and chickadees negotiate around the lethal spines and settle in for a peaceful night of sleep. I am not nearly so adroit and manage to wound myself whenever I work in the roses.

In the sere landscape of late autumn and winter, the rugosa's hips, which resemble brilliant, fairy-sized Chinese lanterns, provide a dependable food source for birds, squirrels, deer, moose, and other mammals. Not only the flesh of the fruit is eaten, but also the hard achenes, or seeds, which despite their unappetizing appearance contain both protein and oils, necessities for hungry winter foragers.

The creatures in my garden aren't the only ones to appreciate the offerings of the rugosas. Early on sunny mornings, I gather a few handfuls of fresh rose petals (never ones that are sprayed with pesticides) and scatter some of them on a shelf in my screened cupboard. After only a couple of days of drying, they are ready to be added to tins of black, green, and mint teas. The scents that waft from my old teapot and the faintly spicy flavor of the teas are delectable.

I use a portion of the fresh flowers in a crock of unsalted butter, alternating a thin layer of petals with about an inch of the butter, until the crock is filled to the rim. This subtle mixture lends a sweet, rosy taste to the breads and muffins I offer my guests. Other petals are added to small antique apothecary jars that I keep in a row on the pantry shelf. Each day I sprinkle in petals and cover them with a dusting of superfine sugar.

Suzy's Sweethearts

'Polareis,' which reminds me of a drowsy baby with a nodding head, has creamy white flowers with pink edges and a pink center. It stands 7-feet tall and thrives in cool weather. Suzy loves the added bonus of the new leaves, which have the scent of sweet grass.

'Jen's Munk' is a Canadian rugosa that quickly becomes established in the garden. The bush is dense with small leaves and forms a great hedge about 5-feet tall. This rose sports clear pink flowers and is "hardly ever out of bloom," says Suzy.

'Rote's Meer' is a German rose that is neat and short, ideal for a small border. Excellent shrub and vigorous, $3\frac{1}{2}$- to 4-feet high. Double, deep purply flowers with stamens visible. This rugosa forms chubby hips, and the fall foliage is colorful.

'Belle Poitevine' flaunts big, blowsy, romantic blossoms. It stands 5- to 6-feet tall and makes a thick hedge or backdrop.

'Charles Albanel' is the solution for rose lovers desiring a carpet of blooms. It is Suzy's shortest pick, standing only $1\frac{1}{2}$-feet tall. This rose "suckers like crazy," and is perfect planted on a steep incline (to help stop erosion). It has large, open, deep pink, and semi-double blossoms. Suzy calls it a "sexy rose, the Marilyn Monroe of the garden."

'Pierette' is the only rugosa with lateral, low-spreading, sideways growth. It stays compact (up to 3 feet) when grown in the sun, but will "reach" with long branches when shaded. The flowers are very large, open, and casual, in the mid-pink range.

'Fru Dagmar Hastrup' sports single, soft pink blossoms with cream-colored stamens. It reaches 4 feet. It has nice big hips, the best fall colors of yellows, oranges, and scarlets.

'Dart's Dash' is for those who love rose hips. It produces the largest, most flamboyant and prolific hips. Deep crimsony purple blooms are "almost doubles, but you can see the beautiful centers."

'Magnifica' has such a pervasive fragrance it penetrates the walls of Suzy's home. "To walk outdoors on a foggy morning is like walking into a cloud of scent," Suzy says. This healthy rugosa has a low, thick growth up to about $4\frac{1}{2}$ feet, and is not prone to stem girdler.

'Schnee Eule' is a pure white compact form—a superb accent rose, perfect for an evening fragrance garden. The white roses contrast with the dark, shiny foliage. Pair it with white nicotiana.

• Suzy's Secrets for Success •

Suzy calls herself a "gardener of least resistance," preferring to use simple methods to keep her rugosa roses healthy and weed free.

Half the battle is won if you site your roses properly in sunny, well-drained soil. They will tolerate high, bright shade, but won't perform as well as those in full sun.

"Roses don't mind being planted deeply," according to Suzy. "They prefer the cooler root zone, and stronger shoots will emerge from beneath the soil."

Use a soil or compost activator to adjust the soil pH; never add lime.

Lay a 16-sheet layer of wet newspaper onto the wet soil surrounding your roses. Top the newspaper with a thick layer of compost. This will encourage beneficial soil microorganisms, help stop soil evaporation, keep roots cool, and exclude weeds.

To sustain the health of your roses, spread compost around established roses and feed occasionally with fish emulsion and seaweed fertilizer. (Suzy depends mostly on compost.)

Although author Henry Mitchell advised pruning these roses while wearing welder's gloves, Suzy says "don't bother." She feels that the beauty of the rugosas is their undemanding nature and advises against pruning too much, "so as not to ruin their character." If you must prune, do it in the spring because fall pruning encourages cane borers; don't cut out canes that are alive.

By the end of summer, the bottles are filled, and I have a good stock of rose sugar for winter and a supply of gifts for my friends. Both of these simple blends absorb the scent and taste of the roses and add an indescribable dimension to many foods.

Perhaps one of the only failings of these roses is the way they blow apart when picked for bouquets; sometimes they drop petals even before they make it into an arrangement. But nothing tops the beauty of a vase with green or autumn-tinted foliage blazing with the smooth, red hips. These striking arrangements last for nearly two weeks indoors, and outside on a protected porch they may survive for a month.

Although the rugosas in my Maine yard are volunteers, the terrace of mixed cultivars of rugosa roses below the pond in my California garden were purposely planted. I wanted an area of shelter for the ground-nesting quail, who are relentlessly hunted by cats; the rugosas, which Gertrude Jekyll describes as having "a ferocious armature of prickles," seemed the perfect solution. I also hoped for a flowery and carefree privacy hedge, or, more poetically, what beloved gardener Rosemary Verey referred to as a "scented screen," to shield a sitting area from a nearby road. The roses are well established now and just beginning to flex their strong, sinewy branches. At night when the 'Blanc Double de Coubert' is in bloom, the dark mound of foliage looks as though it hosts a fleet of schooners with billowing white sails.

I still cannot believe how easily a patch of unruly roses insinuated its way into my heart and changed all my preconceived notions. Before my relationship with rugosas, "rose" seemed like just another four-letter word for "work." Now I know that if you have a little space, sunshine, and enough tolerance to accept the robust habits

of these green heirlooms, you should give them a chance. In the
first season of healthy growth you'll have your own mini-harvests
of scented petals, bouquets of chubby red hips, shelter for the birds,
and a "scented screen" to lure you outdoors and into the pleasures
of your garden.

Travels with Vera

Seeds and leaves,
Roots and flowers,
Ancient plants with
Healing powers.

—SHARON LOVEJOY, *Roots, Shoots, Buckets & Boots*

The insistent, buzzing song of a Bewick's wren and the brilliant spring sunlight awoke me early this morning. I slipped into my robe and headed outdoors to begin my garden chores. Springtime is bittersweet for me. I love these gentle days and their bursting promise of life, but I know that my time in our woodsy California gardens will soon end for this year.

Each May we pack our car and head east to a tiny island off Midcoast Maine. For weeks before we leave, I plant seeds, propagate, transplant, and separate tiny aloe "pups," or offshoots, from their mothers. Many of the seedlings and young plants will remain here, but the long row of pots of Aloe vera atop my potting bench will accompany us on our journey.

It must surprise people to see the rear window of our car snugly packed with over a dozen aloes, from penny-sized ones to the large pots of hen-sized yearlings. Although many stores offer aloe products, I prefer mine fresh and within easy reach. By the time we arrive in Maine, the newly planted youngsters will be firmly rooted and thriving, and the yearlings should be nursing another generation of offshoots.

Aloe, from the Arabic word alloeh for a "bitter, shining substance," is a succulent member of the lily family. The roots of this sculptural African native produce rosettes of pale green lancelike leaves freckled with white and edged with small spines. Unlike the showier members of the lily family, this plant is beloved for its hidden beauty—the healing, cool gel inside each leaf.

Aloe vera sports a host of common names that clearly describe its uses. Known variously as the first-aid plant, the burn plant, the wand of heaven, "the universal panacea," and the plant of immortality, aloe has been a favored healing and cosmetic herb for more than 2,500 years. Historical records claim that Cleopatra attributed her great beauty to her aloe-infused baths and cosmetics.

My Grandfather Clarke, who was a staunch believer in herbal cures for both humans and animals, cultivated my fondness and respect for aloe. In the early 1900s he practiced the art of blacksmithing and horse doctoring. An old leather-bound journal in which Grandfather recorded his herbal recipes contains notes about two of the aloe treatments he formulated for foundering horses. The first was a blend of tincture of aloe, oil of juniper, and oil of sassafras. The second recipe called for a tincture of aloe, oil of juniper, and a quart of raw linseed oil "...to be mixed and applied externally all at once." His spidery notes do not indicate whether his patients lived or perished; but whatever the outcome, his belief in aloe's curative powers lasted his ninety-nine-year lifetime.

Grandfather and his best friend, Bob Lee, tended a battalion of aloes in a motley array of cans and boxes. "This is Mother Earth's cure-all," Bob said earnestly as he dipped a pocket knife into a leaf and dug out some of the bitter goo for me to sample. Neither Bob nor my grandfather had any formal medical training. Instead, they relied on their families' traditional use of the plant and a local Chinese herbalist's recipes. Both men used the soothing aloe gel on their cracked, blistered hands and on minor burns, wounds, and abrasions. A Mason jar of the foul-tasting gel mixed with spring water was kept on a shelf in the refrigerator and sipped (in small doses) for a host of stomach ailments.

I began growing my own aloes when I was a teenager. My first plant produced a litter of babies, and I soon had enough leaves to treat blemishes, a persistent eczema-like rash, and an irritating foot infection. When I was injured in an accident and suffered minor burns on my chin and throat, I consulted a doctor, then turned to my grandfather's favorite plant for relief. Each morning and evening I gently cleansed my skin, slit open a leaf, and applied a thin coating of the gel to my tender face. Despite dire predictions of scarring and redness, my skin healed perfectly in just over a month, and a slight discoloration was all that remained.

"The aloe worked like magic on my burns," I said to Dr. Eugene Zampieron, professor of Botanic Medicine and author of The Natural Medicine Chest. "Aloe has a mixture of antibiotic, astringent, and coagulating agents, and it's a pain and scar inhibitor," he explained.

Grow Aloes on Your Windowsill

I recommend planting aloes in containers that can be easily moved from outside to indoors during cold winters. Make sure your container has a drainage hole; cover the hole with a small piece of screen. This allows water to drain easily, but holds the soil inside the pot. The screen also excludes pesky earwigs, slugs, and sow bugs that may damage your plant.

Pot your aloes in a quick-draining soil (you can buy bagged cactus mix) or mix your own. I follow Liberty Hyde Bailey's recipe for aloe soil, which is three parts sandy loam, one part lime rubble or broken brick (I substitute vermiculite), and a little decayed manure to "strengthen the mixture."

Tamp the soil firmly around your aloe (they tend to topple), and sprinkle a layer of chipped granite or small rocks onto the soil to hold the plant in place and reflect light.

Aloes are drought tolerant, but they cannot abide soggy soil.

To propagate your aloes, strip the small "pups" from the mother plant and put them in a warm, dark place for a week to dry. Tuck the dried aloe into a small pot filled with sandy soil and water sparingly.

Recent studies indicate that it may be used for a number of conditions, from digestive problems to dental surgery, and that one component has the ability to help regulate the immune system and provide anti-viral assistance. "Everyone should grow at least one aloe plant," Dr. Zampieron said emphatically. "During the winter they should be brought indoors and set in a warm, sunny area, then used whenever necessary."

Herbalist Steven Foster, author of *101 Medicinal Herbs*, writes that aloe gel "... relieves pain and inflammation and increases blood supply to injuries by dilating capillaries." He notes that the gel augments tensile strength at the site of the wound and healing activity in the spaces between cells.

Foster warns that some aloe products degrade quickly, and that is precisely why I always travel with my own supply of fresh *Aloe vera*. None of the products I've tested can ever equal the efficacy of these homegrown plants. Whether I am traveling by car or plane, be assured that somewhere nearby lurks a small bag or container filled with a supply of fleshy aloe leaves.

The sun is high overhead as I nestle the final *Aloe vera* into its traveling pot and brush the soil from my nightgown. These pleasurable hours of work guarantee me a steady supply of aloe for our trip east. Whenever calamity strikes, from bites to stings, sunburns to rashes, I'll turn to my faithful aloes for comfort and cures, and perhaps even a refreshing bath Cleopatra-style.

My Grab-Bag Garden

Overwhelmed by choices,
I take the easy way out.

For months, the unruly tower of bulb catalogs beckoned and intimidated me. Everything looked tempting and scrumptious, but the choices were mind-boggling.

I feel that same anxiety when confronted by a bin of perfect tomatoes at our farmers' market. Instead of pondering my choices and inspecting every fruit offered, I implore the grower to rescue me from the dilemma. "Just fill a bag for me," I say with a sense of relief. Oh, that bulb selection could be as easy.

Nineteen years ago, I succumbed to the flashiest new bulb choices available and planted hundreds around my old California cottage. Most of these original settlers survive and thrive to this day. They begin their parade of bloom in mid-winter and flaunt their vibrant colors until summer. But, just as gardens evolve through the years, so has my taste for what belongs in this old-fashioned and wild landscape.

I vowed years ago never to plant anything just for show. Now, I search for rare species and endangered heirloom survivors from centuries past. But there is a catch: Anything I choose must benefit the myriad insects and birds that claim my garden as their home.

One rainy afternoon at our summer cottage in Maine, the winds blew so powerfully across John's Bay that I was forced off my seaside porch and took refuge indoors. The stack of catalogs loomed and leaned precariously over my reading chair; it seemed the perfect time to "weed" through them and place my fall bulb orders.

For an hour, I dog-eared pages with promising choices, then browsed through some trusted garden books in search of information about bulbs that are attractive to pollinating insects. By the end of the afternoon I was at a standstill, with more questions than answers. I felt as though I had stumbled upon an endless line of perfect tomatoes.

My luck changed when what looked like a purple pamphlet, entitled *Old House Gardens*, slid onto my lap and into my life. Never judge a book, or catalog, by its cover. I almost overlooked it, but the question on the cover, "Why Grow Heirloom Bulbs?", intrigued me. Why indeed, I thought, as I read, "They're tough survivors, easy, graceful, wildflowery [love that word], fragrant, unusual, endangered, regionally-adapted, period-appropriate, and just plain gorgeous." From the cover to the last page, I was more than enchanted; I was hooked.

In the middle of the catalog, I found my salvation. "Scott's Garage Sale," the headline read, and went on to explain that founder Scott Kunst—winner of the 2001 Phipp's Conservatory Flora Award—would send me $35 worth of bulbs for $30, if I allowed him to make the choices.

I sat down and jotted Scott a note, asking him to select small, simple bulbs and bulbs whose flowers would offer nectar and pollen for the birds and bugs. Then I requested that they ship my order of two "Garage Sales" at the end of October to coincide with our return to California.

The bulbs arrived before we did. The day we picked up our mail I found a small box with a personal note from Scott, "Thanks for your investment in this experiment with our bulbs.... I'll keep my fingers crossed. I know I love them!" Inside the box were multiple brown bags neatly stapled closed. I was as excited as a child digging into the winning grab-bag offerings at a school carnival.

Nestled inside each sack were innumerable tiny bulbs with lineages to rival a royal family. Jonquil 'Early Louisiana' (1612) rested beside my favorite little daffodil, Narcissus bulbocodium 'Hoop Petticoats' (1629). I set aside 'Twin Sisters' daffodil (1597) and Florentine tulips, which Scott said are scented like violets, to be planted together in a mossy antique pot. As I opened the bags and spread the bulbs across my bench, I read the accompanying labels. Every one contained a bit of fascinating history, a description, and a short primer on care and feeding.

Although I originally hoped to naturalize the bulbs around our pond, a diligent squirrel—who uprooted some of them just minutes after they were planted—helped change my mind. I scoured the potting shed for some of my favorite antique English Sankey terra-cotta pots, filled them with fresh soil, tucked the bulbs inside, and covered the containers with a sheet of chicken wire.

Planting these small, easily overlooked rarities into containers helped me establish an intimate and what I hoped would be a lasting relationship. As each little gem prepared to bloom, I moved the pots (sans the chicken wire) from the workbench and onto the railing and steps of my front porch. Every day brought new discoveries. Bumblebees luxuriated in the flared skirts of golden crocus, beneficial wasps sidled in and out of luminous, dime-sized blossoms, and an unexpected and elusive veil of fragrance greeted visitors at my door.

Underlying my enjoyment of these plants was the feeling of being not only a gardener, but also a curator of sorts. My collection of bulbs linked me to the gardeners who centuries ago tended the same species that now graced my porch. By planting these heirloom bulbs, I might also help, by some small measure, to preserve these ancient and ephemeral green treasures for future generations.

Today I will prepare my new bulb order and intend to follow the coward's course again. I'll order two "Garage Sales," a sampler called "Dutch Crocus Tapestry" (who could resist such a name?), and another sampler called "Easter Basket Hyacinths," which I plan to force on the windowsills of my studio. In the future, Scott will make all the decisions about the bulbs appropriate for my garden. I only wish I could depend on him to choose my tomatoes.

A Sampler of Scott's Favorite Offerings

- **Tulip 'Prince of Austria'** (1860), Zones 4 through 7. "The California condor of bulbs, which almost disappeared from commerce. A strong returner that comes back better than any other tulip; incredibly fragrant, too."

- **Daffodil** (*Narcissus*) 'Conspicuus' (1869), Zones 4 through 7. "Ruffly, pleated rim, bounces in breezes, multiplies vigorously."

- **Lily** (*Lilium*) 'Black Beauty' (new on the block: 1957), Zones 5 through 7. "A deep, dark ruby jewel tone, vigorous and voluptuous."

- **Canna 'Mme. Caseneuve'** (1902), Zones 9 through 11. "Ethereal, mouthwatering, a delicious dessert in apricot and peach with deep bronze leaves. A symbol of the heyday of late-Victorian gardens."

- **Dahlia 'Jersey's Beauty'** (1923), Zones 9 through 11. "Dahlias were brought into gardens by the Aztecs. This one looks like a Miss America in heels. Seven feet tall, palm-sized pink flowers, and goes on a blooming binge in fall."

- **Hyacinth** (*Hyacinthus*) 'Marie' (1860), Zones 5 through 7. "Single, indigo-purple blooms, looks like a roly-poly washerwoman. The oldest commercially available hyacinth, an endangered Victorian survivor."
- **Snake's Head fritillary** (*Fritillaria meleagris*, 1572), Zones 4 through 7. "Grown since before 1800 in America. Quaint, checkered, bell-shaped blooms of rose and pink nod atop 12" stems. A garden oddity."
- **Crocus 'Cloth of Gold'** (1587), Zones 4 through 7. "Once known as the Turkey crocus, this molten golden flower was common in gardens a century ago. Clumps up and covers ground, and is mobbed by bumblebees."
- **Byzantine glad** (*Gladiolus byzantinus*, 1629), Zones 5 through 9. "Brilliant magenta and a quarter the size of other glads. A southern cottage classic."
- **Tuberose** (*Polianthes*) 'Mexican Single' (1530), Zones 9 through 11. "Grow these in pots in your hottest spot, and grow them for their fragrance, not their looks. Common in Williamsburg, Virginia, gardens in the 1730s, and sacred to the Aztecs."

Green Memories

"Having been partly arboreal since the age of eight, I learned early on that trees contain birds' nests, safety, grand vistas, and apples. Climbing tall trees gave me a soaring feeling of achievement... trees were the best toys for hanging, swinging, daring, and showing off."
—BERND HEINRICH, *The Trees in My Forest*, 1998

The trees in my garden are old friends. Like family and trusted confidantes, they are a comforting presence, and I can't imagine living without them. I spent the first nine years of life viewing and puzzling out the complexities of the world from the sheltering branches of an ancient sycamore tree. Whenever I had a problem or needed a place of peace and beauty, I shimmied onto the first swooping limb and crawled slowly up the smooth trunk until I was hidden from view.

It gave me great pleasure to hug that tree and to carry on lengthy, one-sided conversations. My grandmother once told me that a "good listener is a blessing." She assured me that to listen intently to someone is to do them the greatest honor. Well, that tree honored me mightily and helped ease me through a myriad of crises.

I was brokenhearted when my family moved from that old, tree-shaded neighborhood and into a raw Southern California tract. Two skinny lemon trees, hardly big enough to offer shade let alone shelter, were the only things growing in our new yard. In the distance I could see a row of huge eucalyptus trees that were always raucous with crows. The trees stood in a straight line in the median of a busy two-lane road that I was forbidden to cross. For months I only looked at the crow-laden trees until one afternoon, when the feeling of loss and homesickness overpowered me, I climbed our fence and set off for the eucalyptus.

When there was a break in traffic, I ran across the road and stood at the bottom of one of the trees. The first branch was about fifteen feet above me and unreachable.

I started at one end of the median and walked around each tree searching for a branch that would allow an easy access. By the time I reached the last eucalyptus, I knew that the dream of once again losing myself in the familiar sanctuary of a leafy tree was impossible. On that chilly afternoon, permeated with the scent of fallen eucalyptus leaves, I said aloud, "I will always have trees in my life."

Outside my bedroom window, my mother planted an elm tree that was no bigger than my thin ten-year-old arm. I felt hopeless and knew that I would probably be an old, old lady by the time that tree grew to climbable size. "Wherever I live when I grow up," I informed my mom in a disappointed tone, "I will always have trees."

During my adolescence, the need for trees took a backseat to my need for lengthy telephone conversations, immovable beehive hairdos, and carefully plotted strategies to track down my teenage crush, John Horton, in one of his favorite haunts. When I turned seventeen, something wonderful happened that reawakened the primitive, arboreal child inside me.

I met a rosy-cheeked boy named Johnny Arnold. He was from Indiana and had also spent his best times in the branches of a huge old sycamore tree. It was obvious from the beginning that he was my soul mate. We met each afternoon in the limbs of a venerable live-oak tree. We were as drawn to each other and that oak as the crows to the eucalyptus.

Unlike most seventeen-year-olds, our first dates were to nurseries. "I love trees," I told Johnny, and he answered, "I love trees, too." Soon after, we bought a small guava tree and planted it in one of my mom's cracked terra-cotta pots. Nineteenth-century poet Lucy Larcom once wrote, "He who plants a tree plants a hope," and I suppose that was what we were doing. We realized that though we couldn't have an old home surrounded by stately trees, we could plant and nurture small trees and make them an important part of our lives.

A half-hour drive from our neighborhood was a famous Japanese nursery we visited regularly. We learned from the gentle, quiet owner that we could keep our trees happy and healthy by following some of the practices used in the patient art of bonsai. As he spoke with us about the care and maintenance of trees, he walked down a row of plants labeled "Not for Sale." His sun-bronzed hands swept lovingly over his treasured specimens as he told us their names and their incredible ages. "This ginkgo is of ancient lineage, one of the first families of trees to inhabit the earth," he said. He explained that his miniature ginkgo was about seventy years old and was given to him by an uncle who had inherited it from another gardener.

We watched as he turned the shallow, cobalt blue container on its side and used a long knife to slide the ginkgo and soil out of the pot. Slowly and methodically he separated the tiny mass of white roots encircling the tree and trimmed them back by about a quarter. "Never grab your trees by their trunks," he cautioned, "and do all your major root trimming in the autumn when they are dormant." He shook some of the remaining earth from the roots, placed the ginkgo in its original but now uncrowded pot, and filled it with fresh, new soil.

That day we bought a quince, and I tackled my first nerve-racking trimming and repotting. I took care not to tug the tree by its trunk; I gingerly sliced through a solid wall of roots and trimmed judiciously. The quince was just the first of many autumnal groomings, and it not only survived my first awkward-but-loving attempts, it thrived.

Johnny and I began collecting in earnest. We bought a liquidambar tree to commemorate our first anniversary, another guava in honor of my Grandmother Lovejoy, a thimble-sized giant sequoia for my birthday, an Atlantic cedar for Johnny's birthday, and a trio of pencil-thin maples in honor of the birth of our son, Noah. Our trees became our family memory album, but instead of sitting on a shelf and gathering dust, they grew, bloomed, and bore fruit.

• Trees for Life •

- Choose a "memory" tree to honor family members, and pot it in a container twice the width and depth of the root ball. (Make sure your container has drainage holes.)
- Use a copper or aluminum label to note the date, the reason for planting, and the name of the tree.
- Fill the container with bagged potting soil and water thoroughly.
- Trees in containers depend on YOU for survival. Water deeply as needed and feed regularly with a natural liquid fish emulsion-kelp fertilizer.
- To keep them from drying out, some potted trees benefit from a protective coverlet of mulch (keep it away from the trunk). Or plant a cover crop of annuals. (Note: Wilted annuals signal that your tree needs water.)
- Plan to do your trimming and repotting during the dormant period in autumn.

Life changed dramatically through the years, but my love of trees remained as firmly rooted as the sycamore in my grandmother's garden. The tiny specimens that once looked like a collection of toothpicks and popsicle sticks were now robust trees that resided in formidable pots. Whenever Noah and I moved, we always checked out the sunlight and measured to make sure that there was plenty of room for our "friends," the trees. Though we still dreamed of living with trees planted in the ground, trees big enough for climbing, we were content with our mini-forest of memory trees. Fifteen years ago we settled into a small cottage nestled among tall Monterey pines and oaks with branches like long, muscular arms. Our family celebrated the homecoming by planting some native trees in the ground—a first for me—and, in our usual tradition, by nestling a small, spired Alberta spruce in a handsome Italian pot. Just settling the spruce into place made me realize that those first small trees and hopes planted many years ago were finally growing into reality.

It is now my daily joy to make early-morning rounds through the gardens, but I spend the most time tending to (and conversing with) the potted trees that have been a part of my life for over thirty years. Huddled together on a sunny western terrace are the living green pages of our family's memory album.

I have a new tree to plant this week, a diminutive antique apple with pale pink blossoms the color of my new granddaughter's petal-soft cheeks. Another green page, another hope planted, and I will always live with the trees.

Hollyhocks for Fruitfulness

"I want to have a little house
With sunlight on the floor
A chimney with a rosy hearth,
And lilacs by the door;
With windows looking
east and west,
And a crooked apple tree,
And room beside the
garden fence
For hollyhocks to be!"

—NANCY BYRD TURNER

It is the middle of winter, storming up a fare-thee-well, and I can't stop daydreaming about hollyhocks. I'm sitting at the end of our old cherry harvest table, surrounded by packets and bottles filled with hollyhock seeds. The seeds are all gifts, sent to me by fellow gardeners throughout the United States. They're a memory quilt of gardens I have visited and people for whom I have a special fondness.

One packet reads, "clear red singles from Kentucky"; another proclaims, "pale yellow from Santa Fe, N.M." A tag on an old prescription bottle says, "a collection of singles from the Reverend's garden." (I hope they weren't stolen!) A small, crinkled bag bears the message, "These hollyhocks are from the Tor House gardens in Carmel, Calif., compliments of Margot." Tucked into a colorful hollyhock birthday card is a container of the dark seeds with a note that says, "I love hollyhocks, too. I'm sending you a bunch of seeds from my Maine garden. Hollyhocks, as you know, are hardy things that will grow 'most anywhere. Love, Ethel Pochocki."

They will grow "'most anywhere" as long as that "anywhere" is close to humans. Unlike some invasive species from other parts of the world, hollyhocks don't stray far from cultivated gardens. Instead, they linger near our homes, poking through pavement, clinging to fence lines, and peering into cottage windows. Years ago these lanky flowers were famous for flaunting their blossoms in dirty, trash-strewn alleyways. This must be how they earned the unflattering nickname of "alley orchids." I like their other names: "Joseph's staff," an endearment bestowed by the Spaniards; "hockleaf," for their reputed ability to cure a horse's swollen hock (or heel); and "Saint Cutbert's cowl," for their silken collar of petals.

Kathleen Gips' book, Flora's Dictionary, lists the ancient, symbolic meaning of hollyhocks as "fruitfulness." Although they reputedly earned this definition because of their prolific flowering, I think of them as fruitful for the life they attract to a garden.

The old-fashioned single hollyhocks are often the scene of intense duels between hummingbirds and gold-belted bumblebees. Both vie for the nectar in the chalice-shaped blossoms, and the bumblebee, despite its smaller size, usually wins. Painted lady and West Coast lady butterflies flutter up and down the ladder of hollyhock blooms as they deposit their tiny eggs. Checkered skippers and Great Basin white skippers, true to their names, leap and skip among the towering flowers and oviposit (lay eggs) on the hollyhocks' tender new growth.

Hollyhock leaves shot with dozens of holes and peppered with tiny black droppings are a sure sign that hungry caterpillars are in residence. A careful perusal of the crinkled foliage may uncover a dangling chrysalis or the leaf and silken cocoon of a skipper. A

· How to Grow Hollyhocks ·

Hollyhock seeds need light for germination and do best in soil with an average temperature between 65°F and 70°F. You can start them indoors in winter, about eight weeks before planting in your garden.

Sprinkle the seeds on the surface of the soil and pat lightly into place before giving them a gentle sprinkle of water. Seeds take twelve to fourteen days to germinate.

Hollyhocks are fussy about transplanting. Sow them in biodegradable containers that can be put right into the ground.

Once your hollyhocks are a few inches tall, and the soil has warmed, plant them, burying the pot (which will break down, but you may want to slit the sides), 18 inches to 3 feet apart in full sun in rich, loose soil.

Make your hollyhocks happy by keeping them well watered. Put them on a regular feeding schedule, using fish-emulsion or other organic fertilizer.

Do not use insecticides, fungicides, or systemic poisons on hollyhocks. Remember that poisons harm or destroy caterpillars, butterflies, hummingbirds, and bees.

curled leaf swaddled in webbing is a clue that a caterpillar of the painted lady or West Coast lady butterfly rests comfortably inside. They snuggle within the leafy hammock, and sometimes literally eat themselves out of house and home.

Though you may be tempted to tidy up your raggle-taggle patch of hollyhocks after they've bloomed, beware! They are the home, maternity ward, nursery, and restaurant for future generations of skippers and butterflies.

Of all the flowers I mention when lecturing, hollyhocks always elicit the biggest response. Men and women relive their childhood as they share their family recollections and traditions.

Steven Law of Friendship, Maine, told me that he loved to run outdoors early in the morning to pet the bumblebees snoozing inside the hollyhock blossoms. Laurie Archer, who grew up in Danville, Kentucky, and Betty Jackson McKinstry of Kokomo, Indiana, recalled making flower brides from white blooms and bridesmaids from colored blossoms. Isabel Evans of Lake County, California, wrote of her puppet shows with hollyhock ladies strung on wire or mounted on twigs.

My dear friend Millie Baker Stanley reminisced wistfully about her childhood in Ohio in the early 1900s: "We captured fireflies in Mason jars and gently tucked them into open hollyhock flowers on long stalks. Then we knit the blossoms closed with short twigs. During the evening we ran through the gardens using the fairylike hollyhock lanterns to light our way. Of course, we let the fireflies go before we went indoors."

When I was a child, I loved to nibble the plump, green seed capsules of hollyhocks. Traditionally called "cheeses," they do resemble the rounds of cheese you can buy in the gourmet section of most grocery stores. Today when I serve a salad with edible flowers, I stir-fry the "cheeses" and use them instead of croutons. After I toss the salad, I snip fresh hollyhock petals into small pieces and sprinkle them across the top for a finishing confetti of brilliant color.

Laura C. Martin's wonderful book, *Garden Flower Folklore*, mentions that in China hollyhocks are used as ornamentals, as bee plants, and as a food. The coarse, green leaves never topped my list of favorites for a salad, but Laura says they can be cooked as a pot herb and that the young flower buds are considered a delicacy. She mentions that the Romans cooked and ate the greens and used them as an insect repellent. I have serious reservations about serving such a meal to my family. "Please pass the hollyhock stew," I imagine myself saying. "If you don't want to eat it, just apply a thin layer to your skin to repel pests." Noah and Jeff have suffered through my culinary adventures and misadventures, but I think the hollyhock stew/bug repellent might cause them to defect to our neighbor's kitchen.

Anyone who has ever cut a stem of hollyhock has experienced the runny, mucilaginous juice that purportedly is useful in cough syrups and as a poultice for insect bites. Herbalists throughout the ages credit hollyhocks with the ability to kill worms in children, disperse blood clots, act as a diuretic, and cure "humours of the belly." I am not quite sure what those "humours" are, but I am filing away that bit of information. If I am ever diagnosed with that problem, I'll head straight out to the hollyhock border with my snippers and harvest basket.

The weather is letting up a bit now. This afternoon will be a perfect time to sow a few dozen pots of my gift seeds. Though I won't see a green leaf poking through the soil for a couple of weeks, I can already envision our cedar fence graced by a line of swaying hollyhocks. Where there are hollyhocks, there will be fruitfulness. I can almost hear the buzzing of the bumblebees as they dance from flower to flower.

'Tis the Gift to Be Simple

Start a heartfelt tradition by giving trees wrapped in history.

Just about the time the last red leaves fall from the maple trees I begin to suffer my pre-holiday panic attacks. I love the season, but the sheer energy and concentration it takes to plan for the celebrations and gift-giving is sometimes beyond my reach. "Simplify, simplify," wrote naturalist Henry David Thoreau, and during the holidays I chant this mantra as a reminder to pare down and focus on the simplest (but most heartfelt) of traditions and gifts.

A few years ago, I had finished shopping, except for my lifelong gardening friend John Arnold of Albuquerque, New Mexico. He is a passionate cultivator of heirloom bulbs and vegetables, trees, and wildflowers—and an avowed non-consumer. Buying for John is always a challenge, but in the fall of 1999 my problem was solved. Our Maine postmaster, Wayne, had set aside a stack of oversized mail that wouldn't fit into our box. As he handed me the large envelopes and magazines, I spied the words "American Forests Famous & Historic Trees" printed on a fancy catalog which included a selection of heirloom apple trees.

When most gardeners hear the words "apple tree," their thoughts turn to a man called "the tree planter of the millennium," John Chapman, also known as Johnny Appleseed. Chapman left a green wake of apple trees wherever he walked or boated (he packed his seeds under a blanket of moss and mud and transported them on riverways in a double canoe). As he traveled, he purchased large tracts of land, planted orchards, and established nurseries in anticipation of the constant flow of westward-bound settlers. According to author Michael Pollan, by the 1830s Chapman operated a chain of nurseries that ranged from western Pennsylvania into the wild Northwest Territories of Ohio and Indiana.

The Indiana Hoosiers and the Buckeyes of Ohio must have an eternal tug-of-war in regards to John Chapman, who never stayed in one place long enough to sink his proverbial roots (only his trees' roots) into the ground. In lieu of a conventional life, he preferred an outdoor existence and sometimes called a gigantic, hollowed-out sycamore tree "home." Although he lived in both states, by his mid-sixties Indiana had become his headquarters.

A Few Favorite Trees

Historic trees honor the past and are an investment in the future. American Forests not only preserves historic trees, but also plows the profits back into the earth by planting millions of trees in "Global ReLeaf" restoration projects throughout the country.

- **Apple:** Johnny Appleseed's last surviving tree produced the cuttings for the rare 'Rambo' apple (limited number available). Zones 3 to 7.
- **Sycamore:** The Betsy Ross sycamore still shades the Philadelphia home and workshop of the famed seamstress who is credited with creating our stars and stripes. Sycamores are beautiful throughout the year, especially in winter when their subtly hued and mottled exfoliating bark is most noticeable. Zones 3 to 9.

The Indiana connection is what prompted me to order my friend John (who is a transplanted Hoosier) a rare 'Rambo' apple tree, one of the progeny of Johnny Appleseed's last known living specimens. A few years ago the owners of an historic Ohio farm allowed representatives of American Forests to take softwood cuttings from the venerable old survivor, which toppled soon after during a fierce spring storm. Thanks to the owners of the Harvey-Alego farm, the 'Rambo' is now preserved for posterity.

When I ordered the tree for the holidays, the sales representative explained that each specimen is shipped at the time appropriate for

- **White oak:** Abraham Lincoln once addressed a crowd of his supporters in a grove of ancient white oaks in Albion, Illinois. Acorns are now harvested from the original trees and coaxed into life. Plant this one for the next generation. Zones 4 to 8.
- **Tulip poplar:** George Washington planted two tulip poplars at Mount Vernon, his magnificent home overlooking the Potomac River in Virginia. These towering trees now stand nearly 100 feet tall and are pollinated by hand. Imagine owning a tree whose parent was planted by our Founding Father. Other Washington trees available are the Mount Vernon red maple and the George Washington holly, whose branches and berries were used to decorate holiday tables. Zones 4 to 9.

planting in its destination. On gift-giving day, John opened an envelope that contained a certificate of authenticity with a history of the tree, a gift card, and a letter with the approximate shipping date.

In early spring, John arrived home from work and found a long, slender box on his front porch. Inside was a 3 1/2-foot-tall apple sapling; a four-foot greenhouse tube to protect the trunk from scalding sun and hungry rodents; a treated pine support stake; a bird safety net; planting instructions; and a fertilizer tablet. John set to work planting his new tree, not in his hidden backyard, but smack in the middle of his tiny front garden.

· Literary Links ·

These trees are connected to the lives of some of America's most famous writers.

- **Helen Keller** once wrote about feeling the textured bark of a huge water oak in her family's garden in Tuscumbia, Alabama. Acorns from the original tree are harvested from Ivy Green, the family home. These are germinated and the seedlings can be grown in Zones 6 to 10.

- **Harriet Beecher Stowe**, who wrote the history-altering novel *Uncle Tom's Cabin*, lived in Cincinnati, Ohio, in a rambling home flanked by a gigantic native white ash tree which is still thriving and producing offspring for today's gardeners. Zones 4 to 9.

- **Alex Haley**, Pulitzer Prize-winning author of *Roots*, enjoyed the shade of a large silver maple tree at his home in Henning, Tennessee. This native of eastern North America sports silver bark and leaves and has a semi-pendulous form. Zones 4 to 9.

- **American legends**: The Mark Twain bur oak, the Edgar Allan Poe hackberry, and the Henry David Thoreau's Walden Woods red maples round out the offering of literary trees.

Front porches and front gardens are neighborhood builders. John's 'Rambo' tree is now the subject of conversations with passersby and neighbors who pause and chat over the fence. Much as a proud father might show off his newborn, John points out his famous tree and shares its history with interested (and sometimes uninterested) people.

The little apple tree was such a simple gift, but it has given so much pleasure to John and many others. It reminds me of the verse in the old Shaker hymn, "'Tis the Gift to Be Simple." Just change the lyrics a bit and you have the perfect sentiment for the holidays—'Tis the Gift to *Give* Simple.

faretheewell

SELECT BIBLIOGRAPHY

Audubon, John J., and John Bachman. *The Quadrupeds of North America*. Vol 2. Manchester, New Hamsphire: Ayer Company Publishers, 1974.

Bender, Steve, Felder Rushing, and Allen Lacy. *Passalong Plants*. Chapel Hill, North Carolina: University of North Carolina Press, 1993.

Brown, Frank C. *Frank C. Brown Collection of North Carolina Folklore*. Duke University. Durham, North Carolina. 1952–1962.

Comstock, Anna Botsford. *Ways of the Six-Footed*. Ithaca, New York: Cornell University Press, 1977.

Darwin, Charles. *Darwin on Earthworms: The Formation of Vegetable Mould through the Action of Worms*. Ontario, California: Bookworm Publishing Company, 1976.

Foster, Steven. *101 Medicinal Herbs: An Illustrated Guide*. Denver, Colorado: Interweave Press, 1998.

Gertsch, Willis John. *American Spiders*. New York: Van Nostrand Reinhold, 1974.

Gibson, W. Hamilton. *Sharp Eyes: A Rambler's Calendar of Fifty-Two Weeks Among Insects, Birds, and Flowers*. New York: Harper & Brothers, 1891.

Godfrey, Michael A. *A Closer Look*. San Francisco: Sierra Club Books, 1975.

Griffin, Brian. *Humblebee Bumblebee: The Life Story of the Friendly Bumblebees & Their Use by the Backyard Gardener*. Bellingham, Washington: Knox Cellars Publishing, 1997.

Harrison, Hal H. *Eastern Birds' Nests*. Boston: Houghton Mifflin Company, 1975.

Heinrich, Bernd. *The Trees in My Forest*. New York: HarperPerennial, 1998.

Hickman, Mae, Maxine Guy, and Stephen Levine. *Care of the Wild Feathered and Furred: Treating and Feeding Injured Birds and Animals*. Rev. ed. New York: Michael Kesend Publishing, Ltd., 1998.

Hunter, Malcolm L. *Maine Amphibians and Reptiles*. Orono, Maine: University of Maine Press, 1999.

Koch, Maryjo. *Dragonfly Beetle Butterfly Bee*. New York: Smithmark, 1999.

Lipton, James. *An Exaltation of Larks: The Ultimate Edition, More Than 1,000 Terms*. Rev. ed. New York: Viking Press, 1991.

Lorenz, Konrad Z. *King Solomon's Ring*. New York: Thomas Y. Crowell, 1961.

Martin, Laura C. *Garden Flower Folklore*. Reprint. Chester, Connecticut: Globe Pequot Press, 1994.

Martin, Tovah. *Well-Clad Windowsills: Houseplants for Four Exposures*. New York: Macmillan, 1994.

Masumoto, David Mas. *Epitaph for a Peach: Four Seasons on My Family Farm*. San Francisco: HarperSanFrancisco, 1996.

Mathews, F. Schuyler. *Field Book of Wild Birds and Their Music*. New York: G. P. Putnam's Sons, 1904.

Meeuse, B. J. D. *The Story of Pollination*. New York: Ronald Press, 1961.

Milne, Lorus J. *The National Audubon Society Field Guide to North American Insects & Spiders*. New York: Knopf, 1980.

Nabhan, Gary Paul. *Cultures of Habitat: On Nature, Culture, and Story*. New York: Counterpoint Press, 1997.

Newman, Leonard Hugh. *Create a Butterfly Garden*. London: The Country Book Club, 1968.

Olkowski, William, Helga Olkowski, and Sheila Daar. *The Gardener's Guide to Common-Sense Pest Control*. Newtown, Connecticut: Taunton Press, 1996.

Pyle, Robert Michael. *Audubon Society Handbook for Butterfly Watchers*. New York: Scribner, 1984.

Rennie, James. *Insect Architecture*. Reprint. London: John Murray, 1957.

Sandbeck, Ellen. *Slug Bread and Beheaded Thistles: Amusing and Useful Techniques for Nontoxic Housekeeping and Gardening*. Reprint. New York: Broadway Books, 2000.

Terres, John K. *Songbirds in Your Garden*. Algonquin Books: Chapel Hill, North Carolina, 1994.

Thaxter, Celia. *An Island Garden*. Reprint, Boston: Houghton Mifflin Company, 2002.

Thoreau, Henry David. *Walden* and *Civil Disobedience*. New York: Signet, 1942.

Trafton, Gilbert H. *Bird Friends*. Boston: Houghton Mifflin Company, 1916.

Tweit, Susan J. *Seasons in the Desert: A Naturalist's Notebook*. San Francisco: Chronicle Books, 1998.

Webb, Mary. *Precious Bane*. Notre Dame, Indiana: University of Notre Dame Press, 1980.

White, E. B. *Charlotte's Web*. New York: HarperCollins, 1952.

Wright, Amy Bartlett. *Peterson First Guide to Caterpillars*. Boston: Houghton Mifflin Company, 1998.

Zampieron, Eugene R. and Ellen Kamhi. *The Natural Medicine Chest: Natural Medicines to Keep You and Your Family Thriving Into the Next Millenium*. New York: M. Evans & Co., 1999.

INDEX